The Innovation Age

A New Perspective
on the
Telecom Revolution

The Innovation Age

A New Perspective
on the
Telecom Revolution

Peter K. Pitsch

Hudson Institute
The Progress & Freedom Foundation

For Geoffrey, Madeleine, and A.M.D.G.

Hudson Institute
The Progress & Freedom Foundation

ISBN 1-55813-059-4

Printed in the United States of America

This book may be ordered from:

Hudson Institute
Herman Kahn Center
P.O. Box 26-919
Indianapolis, Indiana 46226
(317) 545-1000

The Progress & Freedom Foundation
1301 K Street, N.W.
Suite 650 West
Washington, D.C. 20005
(202) 289-8928

Contents

Acknowledgments

A book is not written so much as it is produced. I wish to thank Hudson Institute and The Progress & Freedom Foundation for their extensive support. Many thanks to Northern Telecom, Inc. for their generous funding of this important project. Kudos to David Murray and David Teolis with whom I collaborated on earlier papers and whose ideas and in some cases prose are evident in my discussions of the computer industry and universal service. Also, valuable comments on earlier drafts were made by Sam Karnick, David Murray, James Gattuso, Evan Kwerel, Tom Duesterberg, Quincy Rodgers, Richard O'Donnell, Kent Lassman, Bill Doying, and Thomas Spavins. Cynthia Aungst, Robert Champ, and Julianne Lee were most helpful in the production and copyediting of the book. I wish to recognize my greatest debt to Margery Erickson who from start to finish was instrumental to the writing and researching of this book. Lastly, I note that some of the analysis herein is based on work done on behalf of telecommunications clients. In all cases the views expressed and any errors are my own.

Foreword

The increasing pace of change is a defining feature of our time. Far more than those who lived during the Renaissance or the Industrial Revolution, we are self-consciously aware of the transcendent characteristic of this period in history: we truly live in an age of innovation. Even the most conscientious technophobes are finding it difficult to ignore the waves of technological change that are rolling through the global economy.

As a member of Congress whose chief legislative and oversight responsibilities cover telecommunications, I have come to see just how outdated our regulatory and legal structures are in relation to the world of technology that they are purportedly designed to rationalize. We need to rethink our application of old-style bureaucratic regulation to today's fast-changing, high tech industries. That is because delay, which typifies regulatory bureaucracies, is utterly incompatible with the pace of new knowledge and invention. An entirely new political economy of innovation is required.

In *The Innovation Age,* Peter Pitsch, a communications law expert and former senior official at the Federal Communications Commission, has provided policymakers and general readers a much-needed new perspective on the regulation of technological change. He starts by applying the insights of two great Austrian economists, Joseph Schumpeter and Friedrich von Hayek—both of whom were principally concerned with market processes and the dynamics of change. Schumpeter stressed the importance of free enterprise in creating breakthrough innovations—a notion captured in his now famous phrase "creative destruction." Hayek observed the astounding ability of decentralized markets to rationalize complexity and achieve what he called "spontaneous order."

The Innovation Age combines these two concepts into one central insight: the need for markets in discovering, rationalizing, and evaluating innovation—big and small—has assumed new importance in an era of increasing change and technical complexity. In Pitsch's phrase, we must therefore concern ourselves with how regulation affects the market's ability to act as a "discovery process."

There is very real urgency to this conclusion. While *The Innovation Age* examines the telecommunications market as a vehicle for illustrating the inadequacy of current government policy, the topic is hardly academic. Unless the wisdom Pitsch imparts is adopted by lawmakers and regulators, the expected dramatic improvements in telecommunications services will be lost—and with them the jobs, increased commerce, heightened productivity, and higher standards of living we all hope to enjoy.

Telecommunications industry news is ubiquitous. But despite the seeming oversupply of market information, it remains difficult to predict which new telecommunications services will be successful. Pitsch demonstrates that sensible predictions nonetheless can be made, based on the enormous cost and performance improvements in telecommunications services overall. He makes cogent "Schumpeterian bets" on the effect of the telecommunications revolution in specific areas such as intellectual property, subscription programming, and advertising. His best bet of all, though, is that systemic and pervasive improvements will occur in the market's own "discovery process."

Pitsch also addresses the many important questions for federal and state regulators left open by the Telecommunications Act of 1996. Federal regulation, he correctly argues, should be designed to promote instead of hinder competition. Antitrust enforcement should be conducted on a case-by-case basis. Technical standards should be set by private parties in the marketplace. Telephone subsidies should be vastly reduced, targeting them to the people who truly need them. Expansion of so-called "universal service" is premature: until we learn what new technologies and services can be made available, the demand for them, and their cost, it is impossible to judge what the analogue of basic telephone service will be in the future. And the current system of spectrum management should give way to privatization, because the FCC necessarily lacks the information and incentives to make correct decisions.

Telecom improvements are a crucial factor driving the entire technological revolution of our time, because they accelerate the beneficial exchange of ideas and sharpen the market's ability to discover new, more efficient arrangements. Pitsch demonstrates that continuing technological advance, increased competition, and deregulation will further propel the telecommunications revolution. He explains why recent experience with the unregulated computer industry provides a compelling policy paradigm for understanding the rapid evolution of the telecommunications industry.

In five years, most of the myriad histories and accounts of the telecommunications revolution will be irrelevant. In a concise, illuminating form, *The Innovation Age* provides needed perspective on what's important and likely to be enduring about this revolution, and how policymakers should act to nurture it.

CHRISTOPHER COX
U.S. REPRESENTATIVE

Preface

Innovation has become the byword of the era. Each day brings new developments in technology and information exchange, engendering sweeping, progressive changes in the ways we conduct our economic, social, cultural, and even political activities. The fast and continuously accelerating rate of this transformation, across all sectors of society, has brought about a third great era of human history, commonly termed "The Information Age." Though information is indeed the vehicle by which these unprecedented changes are being wrought, change itself is now such a universal constant that our times might be more appropriately termed, in Peter Pitsch's words, "The Innovation Age."

The telecommunications industry practically defines our era, with its phenomenal rate of growth and innovation. Events transpire at a speed far outstripping governments' ability to generate regulations or even to foresee instances in which public interest might be warranted. In *The Innovation Age*, Peter Pitsch shows that the United States cannot maintain its preeminent position in information technology in an increasingly competitive global economy unless a number of outmoded regulatory restraints are significantly altered or removed altogether. "Universal service" goals from the telephone era, for example, are inappropriate to the new technologies, especially when it is impossible to foresee just what "basic telecommunications service" might comprise a year or two into the future. Yet governments still try to impose these obsolete rules on new technologies, a quixotic quest that only succeeds in decreasing competition, increasing prices, and impeding innovation.

The Innovation Age documents how artificial regulations on markets hinder the very progress and productivity they were designed to stimulate. Pitsch uses the computer industry as an example of successful "benign neglect" on the part of the federal government and cogently illustrates that the path to an "Innovation Age" should be kept clear of ill-fitting limitations and rules, however well-meant. Pitsch asserts the superiority of a market-based approach to regulation, deriving his perspective from the ideas of the visionary Austrian economists Von Hayek and Schumpeter. Their analyses of the shortcomings of planned economies engendered the insightful concepts of "creative destruction" and "spontaneous order," and Pitsch shows that these ideas are superbly applicable to our current technological situation, noting where state and federal regulation efforts could be refocused to encourage competition and reap the benefits of the "discovery process."

Most importantly, *The Innovation Age* paints a bright picture of the future through an open-road policy toward the telecom industry. Improvements in telecommunications will increase the nation's collective capacity to conduct vital transactions, transmit ideas, and maximize productivity while economizing time. In a world where time is increasingly crucial, *The Innovation Age* embodies a telecommunications policy approach whose time as definitely arrived.

Many generous contributors to Hudson Institute and The Progress & Freedom Foundation have helped make this book possible, including Northern Telecom, Inc. Although these organizations bear no responsibility for the ideas and conclusions of this work, Hudson and The Progress & Freedom Foundation gratefully acknowledge their support.

HUDSON INSTITUTE
THE PROGRESS & FREEDOM FOUNDATION

Chapter One
Innovation Age

Age-Old Lessons

On October 21, 1805, British ships under the command of Admiral Horatio Nelson defeated French and Spanish fleets off Cape Trafalgar in the south of Spain, decisively ending Napoleon's eight-year attempt to gain the command of the sea he needed to attempt an invasion of England. Invasion by the fleet of flat-bottomed boats assembled by Napoleon on the other side of the Straits of Dover had proved impossible: the British navy had stood in the way. For more than two years, English frigates had constantly patrolled just outside ports on the French side of the Channel. Through constant practice their captains and crews had become highly skilled seamen. As First Lord of the Admiralty St. Vincent put it: "I do not say the French cannot come, I only say they cannot come by sea."[1]

St. Vincent's boast was bolder than he knew. Historian David Horwath has pointed out that a relatively few steamboats could have been used as tugboats to tow Napoleon's troop-carrying flat-bottomed boats across the channel. On a windless and foggy night the English warships would have been helpless to stop them. This scenario is not far-fetched. Since 1801 a steamboat named the *Charlotte Dundas* had successfully been pulling barges along the Forth and Clyde Canal in Scotland. Also, Robert Fulton's steamer had been built in Paris in 1803. Within four years of Trafalgar, an open sea passage was made by a steamboat in America.[2]

This historical "might-have-been" illustrates two lessons about innovation. First, innovation has long, if not always, been critical to human affairs. Horwath's proposed use of steamboats might have changed the course of English and possibly world history. If Napoleon's formidable army could have safely crossed the Channel, who can be sure what would have happened? Second, the uses to which new technologies can be put are only dimly appreciated in their early years, with their ultimate impact being highly unpredictable. The English did consider the possibility that Napoleon might find an alternative means of ferrying his troops across the Channel. There were wild rumors about Napoleon's troops coming via a Channel tunnel, balloons, or huge rafts powered by windmills—but all the while the potential of steamboats was overlooked.[3]

These lessons on the nature of innovation have assumed enormous importance today, for the pace of innovation has accelerated considerably in the two hundred years since 1805. House Speaker Newt Gingrich and Vice President Al Gore are among many who frequently remark on the technological changes they see transforming our economy. Revolutionary developments in computer and fiber optic technologies, in particular, are cited as having radically altered the way we process and communicate information today.

These advances in digital technologies have sparked a new awareness of information as central to economic activity and produced a widespread belief that the world is entering a new "Information Age." As consumers and businesses adjust to the new opportunities created by digital technologies, they are transforming the economy. Alvin and Heidi Toffler characterize these developments as "the third wave" of human civilization, promising change comparable in importance to man's adoption of agriculture eight thousand years ago ("the first wave") and the industrial revolution in Europe and the U.S. in the 19th century ("the second wave"). The Tofflers predict that the means of production, distribution, marketing, entertainment, and communication will become "deconcentrated" as digital advances reduce the importance of economies of scale and lower the cost of decentralized and diversified approaches.

The societal transformations now occurring are profound and information-driven, yet it would be a mistake to view them as discrete, predictable, or one-time phenomena. Rather, the defining feature of this new era will be continuing and accelerating change. At the turn of the century, a country with an economy growing at 4 percent per year would have been considered to have a high growth rate. Now to rank in the top tier of growing economies, a growth rate closer to 10 percent is necessary.[4]

To understand what is happening in today's world economy, we need to look beyond the development of particular technologies to the change in the pace of change itself. Thus, the Information Age could more appropriately be termed the "Innovation Age." The importance and unpredictability of innovation—the age-old lessons—have mushroomed as the speed of technical change has quickened. Modern economies will become increasingly dynamic, unpredictable, and complex. The potential gain from innovation will be both greater and less predictable, and the unforeseen effects of regulating economic systems will become more significant and costly. This new age of innovation requires us to rethink our approach to regulation and the role of government generally.

Nowhere are these lessons more applicable than in telecommunications—the networks that allow huge quantities of information to be sent and received at high speed. These networks are one of the key forces behind the heightened pace of innovation. Experience with the computer industry—which, as we will see, is intimately related to the telecommunications industry—suggests that the magnitude of the ongoing "telecom" changes will be enormous and unpredictable. Analyzing the telecom revolution within the framework of the Innovation Age will help us make some basic predictions about the revolution as well as guide policymakers as they change specific regulations in the hopes of benefiting consumers.

Innovation Prophets

Writing in the first half of this century, two Austrian-trained economists—Friedrich August von Hayek and Joseph Alois Schumpeter—changed the way we think about the role of complexity and innovation in economic systems. Born in Vienna in 1899, Hayek was professor of economics at the Austrian Institute of Economics Research, the London School of Economics, the University of Chicago, and the University of Freiburg. Schumpeter was born in 1883 near Jihlava in what is now the Czech Republic. He trained at the University of Vienna and joined the faculty of Harvard University in

1932. These two economists' legacy has been their insights into the workings of the market and its comparative advantages over socialism and central planning regimes. In today's increasingly complex, fast-changing global economy, their analyses are even more compelling than when they were first written. Their insights provide a basis for understanding our future economy and the government's role in it. They have become economic prophets for the Innovation Age.

Spontaneous Order

Friedrich Hayek is best known for his 1944 classic, *The Road to Serfdom,* in which he warns that the loss of economic freedom eventually leads to the loss of civil liberty as well. Perhaps his most important work, however, was his discussion of the nature and operation of markets in *The Constitution of Liberty* and related philosophical essays. In these writings, Hayek expounds on how effective human institutions such as markets develop spontaneously, without the benefit of any overt planning. Indeed, Hayek contends that markets work far better than "planned economies" because they "utilize the knowledge and skill of all members of society to a much greater extent than would be possible in any order created by central direction."[5]

One of the best illustrations of this phenomenon, according to Hayek, is the price mechanism. In digital parlance, millions of firms and consumers "parallel-process" their economic plans around the available price data. Economic theory has developed an impressive system of equations showing how prices are determined by the desires, resources, and knowledge of all the participants in a market. But Hayek contends that it is absurd to think that anyone could ever ascertain all the data needed to use those mathematical equations to arrive at the numerical determination of prices. He warns against such intellectual hubris:

> [The] purpose [of the mathematical equations] is exclusively to describe the general character of the order that will form itself. Since this order implies the existence of certain relations between the elements, and the actual presence or absence of such relations can be ascertained, the prediction of such an order can be shown to be false, and the theory will

> thus be empirically testable. But we shall always be able to predict only the general character of the order and not its detail.[6]

In emphasizing the enormous complexity that lies beneath the market system, Hayek is advising government policymakers to be cautious in their attempts to redesign it.[7] He says, "It is high time . . . that we take our ignorance more seriously."[8]

Creative Destruction

Joseph Schumpeter, like Hayek, recognized the complexity of the economic process; he believed capitalism to be superior to socialism and was leery of regulatory tinkering. In 1942, while an economics professor at Harvard University, he published *Capitalism, Socialism, and Democracy*, the culmination of four decades of thought. In that work he characterizes the market process as an organic one in which considerable time is required for the separate elements to reveal their true features and ultimate effects.[9] For Schumpeter, however, the defining feature and the key to understanding the fundamental workings and benefits of capitalism is its dynamism:

> The fundamental impulse that sets and keeps the capitalist engine in motion comes from the new consumers' goods, the new methods of production or transportation, the new markets, the new forms of industrial organization that capitalist enterprise creates . . . the same process of industrial mutation—if I may use that biological term—that incessantly revolutionizes the economic structure from within, incessantly destroying the old one, incessantly creating a new one. This process of Creative Destruction is the essential fact about capitalism.[10]

Schumpeter's creative destruction hardly makes for a complacent existence for the participants. Businesses are enticed to compete vigorously for "spectacular prizes," even though in the end most will receive only modest compensation.[11] It is the resulting creative destruction that "in the long run expands output and brings down prices," but that conclusion is reached through a "competition which commands a decisive cost or quality advantage and which strikes not at the margins of the profits and outputs of the existing firms but at their foundations and their very lives."[12]

Schumpeter noted that this competitive behavior is manifested in improvements to quality and performance rather than simply in lower prices. In other words, competitive improvements in technology and production, despite the fact that some firms enjoy temporary monopolies or market dominance, will benefit the average consumer. Schumpeter was also keenly aware that market behavior at any given moment provides only a misleading snapshot of the economy, not the sort of moving picture that conveys the gains and losses to consumers.[13]

Schumpeter believed that the benefits of these successive innovations are so great that they outweigh the economic drag created by the temporary monopoly that a firm deploying a new innovation may enjoy. He looked at the technical innovations of his day and concluded that the cost reductions from farm machinery, steel mills, and airplanes dwarfed the static losses that arose because the industries that produced these modern wonders did not meet the conditions of perfect competition.[14] Moreover, he realized, the nature of most monopolies operates to reduce the losses predicted by the static monopoly model:

> It is hardly necessary to point out that competition of the kind we now have in mind acts not only when in being but also when it is merely an ever-present threat. It disciplines before it attacks.[15]

Indeed, Schumpeter concluded, departures from the conditions of perfect competition may in many instances benefit consumers and regulators must therefore exercise great care in attempting to remedy them.[16]

Compelling Changes

Although Hayek and Schumpeter are familiar names in the pantheon of economists, their insights all too often have not been given the weight they deserve. Perhaps this is understandable, since Hayek's and Schumpeter's views bring into question the usefulness of much of the static efficiency analysis that is the main subject of contemporary economic textbooks.[17] Nonetheless, both men recognized long ago the importance of discovery to the workings of an economic system. Hayek's spontaneous order and Schumpeter's creative destruction together expose the true comparative advantage of the marketplace: it is a better *discovery-process*. It is better especially at discovering order and creativity. Economist Thomas Sowell, who has written extensively on the operation and comparative efficacy of various decision-making processes, notes that the importance of the market's ability to "discover" is more important when more is unknown—that is, when either the relative worth of existing goods has changed or new opportunities are now within reach. He observes that

> the role of prices as transmitters of knowledge is more readily seen in a changing economy than in a static one. If the economy maintained the same technological capabilities at all times, and tastes were unchanging and population size stationary, one way or another most of the essential knowledge about how things ought to be produced would eventually percolate through the system. When all these (and other) variables are changing constantly, however, the knowledge problems become staggering—if viewed from the standpoint of a given individual trying to understand it all.[18]

The key advantage of the marketplace as a discovery-process is that it effectively conveys new information and creates the incentive to act on it for all participants throughout the economy. Sowell writes as follows:

> [F]or example, [if] new deposits of iron ore are discovered at a time when there is a growing demand for office furniture and a declining supply of trees, all that the stores that sell office furniture need to know is that the wholesale price of steel desks, tables, and cabinets is falling relative to the wholesale price of the same items made of wood. They may do no

> more than reflect these relative price relationships in the retail prices they charge for steel and wooden office furniture. Those consumers who absolutely swear by either steel or wooden office furniture may just continue their respective preferences, but others, who are either more flexible or more pressed for cash, will tend to substitute the material that is getting cheaper for the material that is growing more expensive. The net result is that the economy as a whole incrementally substitutes the material that is becoming more abundant for the material that is becoming more scarce, without either the consumers, the retailers, or even the wholesalers necessarily understanding why prices are changing the way they are.[19]

Several widely noted economic changes are now making Hayek's and Schumpeter's insights into the market as a discovery-process more compelling with the passing of every day. The globalization of markets has added a new dimension to an already complex business environment. The collapse of communism in the Soviet bloc countries and the reduction of tax, tariff, and other barriers to foreign investment in many of the so-called developing countries promise to increase economic growth rates dramatically in these areas of the globe.[20] The reduction of trade barriers, lower transportation costs, and the growth of English as the preferred language of commerce have expanded markets and increased the scope of specialization and the intensity of competition throughout the world. Two frequently cited examples of these new conditions are that "financial markets are becoming borderless" and "the workforce is becoming global."[21]

Likewise, technological advances are making the world a far more dynamic place. Uncertainty and opportunity have increased as the pace of innovation has accelerated across a broad range of technical fronts. Experience suggests that accelerating change will increase uncertainty because bigger changes require more adjustment time to reach equilibrium. The dramatic price and quality changes that innovation causes will produce unanticipated and delayed reactions as businesses and consumers learn how to exploit and adapt to these changes. Unconventional ideas take longer to develop. For example, personal computers were initially used as a more efficient way to type letters; only later did they facilitate the flattening of corporate hierarchies as the "networked workplace" eliminated the need for many intermediaries.

In the digital field alone, continuing major advances can be expected in semiconductors, parallel processing, and optoelectronics: three leading technologies driving computer and telecom performance. Early next century, semiconductor improvements will allow the production of microprocessors having the speed of today's entry-level supercomputers. At the same time, parallel processing, which enables multiple processors to divvy up tasks, will become pervasive—perhaps even linking multiple microprocessors on a single chip. Advances in optoelectronics—the lasers and optical switches that make fiber optic cables possible—have made 10,000-fold improvements over copper-wire-based telecom systems. In the future, fiber optic systems are expected to improve from their current "throughput" of 1 or 2 gigabits of information per second to 10 and eventually to 100 gigabits per second.[22] Similar improvements are expected in storage, programming, speech recognition, compression techniques, and wireless technologies. Such a broad range of change among competing and complementary technologies can be expected to create numerous unanticipated interactions.

In addition, technological change has elevated the importance of Schumpeter's notion of creative destruction.[23] As we will see later, the overthrow of IBM's "big iron" mainframe computers by the personal computer is a spectacular example of creative destruction. Microsoft, Lotus, Novell, Apple, Compaq, Dell, Intel, and the other companies that "destroyed" the mainframe market at the same time created enormous benefits for the consumer and business users of the personal computer. Further, the faster pace of technological change has reduced the retarding effect that a dominant firm's embedded base might have on its incentive to innovate. In the past, innovation often came from "outsider" firms.[24] In today's world of foreshortened product cycles, business managers cannot afford to rest on their laurels. Andrew Groves, the CEO of market leader Intel, has explained his competitive strategy with a revealing quip: "Only the paranoid survive."[25]

Lamarckian Innovation

The compelling nature of the Hayekian and Schumpeterian perspectives in the Innovation Age is well illustrated by Michael Rothschild's comparison of the modern economy to an ecosystem.[26] In his book *Bionomics*, Rothschild develops detailed analogies between biological systems and economic activity. Like Hayek, Rothschild sees economies as complex, spontaneous orders that no one has designed or is capable of designing. Rothschild is also critical of traditional economic modes of analysis because they virtually ignore the role of growth and change in economies. In his view, modern economic theory sees the economy as if it were a machine modeled on Newtonian mechanics; in reality, says Rothschild, economies are much more like ecosystems and better explained by Charles Darwin's concepts of evolution and natural selection. Machines do not change or grow on their own. Economies, on the other hand, are in a constant state of change and are continually adapting to internal and external factors.

As Rothschild recognizes, moreover, the ecosystem analogy fails to capture one important feature of economic change: While organisms in an ecosystem change through relatively rare gene mutation, economic participants can change through their own efforts. In this regard, economies better fit Jean-Baptiste Lamarck's mistaken view of evolution. Lamarck theorized that animals passed on acquired traits to their offspring. Under this theory, the giraffe developed a long neck because giraffes that "stretched" their necks to reach ever-higher leaves passed on this trait to their descendants.

The Lamarckian mode of change speeds up the rate of economic advance because this kind of change does not have to wait for beneficial genetic mutations to occur. As Rothschild observes, economic change stems largely from recorded information that now exists outside human DNA. Unlike an ecosystem, the growth in this "corpus" of information is not limited by the slow pace of chromosomal change. Because the growth of information is not so limited and economic participants can change through their own efforts, the rate of economic growth can actually accelerate over time as participants react to one another's new ideas.

As Hayek and Schumpeter realized, the market fosters economic development by discovering innovation and spreading its effects throughout the economy through voluntary for-profit transactions. Oftentimes, the change will simply be an idea and have the attributes of what economists call a "public" good. (In other words, once an idea is learned, it costs little to pass on and will be virtually free to others.) Many innovations—such as the "supermarket," minivan, facsimile machine, and cordless phone—were quickly copied at little or no cost using existing technologies. Moreover, even where the users of new technology must compensate its creator, they often pay far less than it cost to produce. In these ways innovation can supercharge economic growth. The work of several economists inspired by Joseph Schumpeter, especially Paul Romer of the University of California at Berkeley, supports this view. Romer has developed what he calls the "New Growth Theory," a central tenet of which is that there is enormous scope for the development of new ideas.[27]

A variety of factors can spur innovation and therefore economic growth—for example, greater research and development, more specialization, better communications, lower transportation costs, deregulation, privatization, and reduced governmental barriers to trade. Importantly, in addition to increasing innovation directly, many of these factors—such as free trade and improvements in communications and transportation—indirectly expand the universe of knowledge by increasing the number of beneficial interactions between new ideas and innovative developments. For example, economists Gene Grossman of Princeton University and Elhanen Helpman of Tel Aviv University theorize that it is the rapid dissemination of technology and industrial ideas made possible by eliminating trade barriers that produces free trade's dramatic improvements in economic growth.[28]

The potency of this "interaction effect" can be illustrated by an everyday example. Consider five people shaking hands. If each person shakes hands with every other person, there will be 10 handshakes or "interactions." This result is given by the formula $n(n-1)/2$, where n is the number of people shaking hands (e.g., 5x4/2=10). If the number of people shaking hands increases to 6, the number of handshakes increases from 10 to 15 (6x5/2=15). It can be shown algebraically that increasing n by 1 will produce an n increase in the number of handshakes (e.g., where n=5, increasing n by 1 will produce 5 more handshakes). Thus, the greater the number of interactors (n) to start with, the greater the resulting increase in interactions (n)

will be. Analogously, economic factors such as better communications, transportation, and trade can be expected to increase the number of interactions (n). Each continuing improvement in these factors will then produce an effect that builds from a bigger base. Together, the cumulative impact of these factors will be a geometric increase in the number of beneficial interactions among innovative developments.

New Vision

Humble Futurists

The previous analysis of the changing nature of our economy requires that futurists in the Innovation Age undertake their trade with humility. The growing complexity and uncertainty in the economy are making an already tough job ever harder. On the one hand, this growing complexity makes it increasingly difficult to make technological predictions, because out of the myriad possibilities we do not know which particular technical modalities will become state-of-the-art or what real improvements they will generate. On the other hand, it has become more and more likely that there will be radical changes in the technical status quo.

Schumpeter would warn us that explosive change in some shape or form is more likely than ever. Predicting that the status quo will continue is therefore not a safe course. This notion is borne out by the rapid increase in economic growth rates. Beginning in 1780 it took Britain sixty years to double its output per person. Beginning in 1880, Japan achieved this feat in thirty-four years. Beginning in 1966, South Korea did it in eleven years.[29]

Hayek, in turn, would warn against the conceit of "knowing" more than we can know. Simplified or "hero-based" analyses almost certainly will be too simplistic. Often, market processes mask the fact that a wealth of possibilities and trade-offs exists. Nobel laureate Robert Fogel's analysis of the early impact of American railroads illustrates this point.[30] Fogel found that even the railroad was not vital to the economic growth that occurred during the nineteenth century. Despite the railroad's rapid growth, eventual dominance of inland transportation, and enormous use of capital, it did not make an overwhelming contribution to the productive potential of the economy. Canals and other forms of water transportation often provided an alternative means of shipping during this period, and in the absence of the railroad they would have been economically feasible in the north central states and Texas. Fogel estimates that the social savings created by railroads in transporting all commodities was well below 5 percent of gross national product. Fogel also found that the derived effect of railroads on iron production and other industries was often overstated. His analysis shows that economic growth was a consequence of numerous innovations developed over the course of several centuries that were applied to a broad spectrum of economic processes. It is evident that the complexity of the growth process, even in a much slower paced economy, makes understanding the impact of particular technologies extremely difficult.

Reticent Regulators

These lessons have serious ramifications for regulatory policy. What economists would call the political economy of innovation must undergo a fundamental reassessment. The classic justification for utility regulation in this country has been the need to control economic power in industries that would most likely be dominated by one firm, e.g., railroading, telephone service, electric power, and natural gas distribution. Regulation has also been used to check the competitive process in less concentrated markets such as trucking and inland water shipping.[31] Today, growing global competition, merging markets, and technological advances have undercut these rationales and made the unintended and untoward effects of government regulation far more costly than in the past. Moreover, the accelerating pace of change makes regulatory costs increasingly unacceptable. The "static" efficiency gains from price regulation of a firm with market power can quickly be swamped by the foregone benefits of innovation due to the delay that often attends such regulation.

Increasing complexity and accelerating technological advances should change how we approach regulation. First, the daunting web of interconnections that now underlies the "spontaneous order" should give policymakers pause as they seek to "improve" on the marketplace. This is especially true in those many instances in which regulation is intended to protect the status quo from technological innovation. The substantial efficiency gains from innovation will likely be more diffuse and even harder to trace than they were in the past, although the loss from job displacement will be concrete and immediate. While the short-term political appeal of giving in to this temptation will probably increase, we must recognize that the costs of protecting the status quo from changes in technology will increase even more dramatically.

Second, the faster the potential rate of innovation, the more important dynamic considerations become and the greater the care that must be taken to assure that regulations designed to address efficiency concerns do not have the unintended effect of retarding innovation. In Schumpeter's words,

> A system—any system, economic or other—that at every given point of time fully utilizes its possibilities to the best advantage may yet in the long run be inferior to a system that does so at no given point of time, because the latter's failure to do so may be a condition for the level or speed of long-run performance.[32]

This statement is true because of the impact of compounding a series of small improvements in performance over a sustained period. Consider a particular regulation of an industry that can achieve efficiency gains of 10 percent from lower prices, but only at some cost in slower innovation and growth. Assume that under the regulation the rate of growth in that industry's value added to society (i.e., its contribution to consumer and producer surplus) is 2 percent per year, and that without the regulation the rate would be 2.5 percent. In that case, as can be seen in the table below, the costs of foregone innovation begin to exceed the benefits of price regulation in twenty-two years. If the unregulated annual growth rate is assumed to be 4 percent, however, the crossover would occur in just over five years.

TABLE 1
The Power of Growth

Scenario	Static Efficiency Gain from Regulation	Growth Rate Under Regulation	Growth Rate Without Regulation	Cross-over Point	Net Gain Without Regulation after 20 years
	(%)	(%)	(%)	(years)	(%)
1	10.0	2.0	2.5	21.6	—
2	10.0	2.0	4.0	5.4	32.7
3	5.0	2.0	3.0	5.3	15.5
4	1.0	2.0	3.0	1.0	21.3
5	1.0	2.0	4.0	.52	46.0

The table also shows that if static efficiency gains from regulation are smaller (which is far more likely),[33] the point at which the performance of the industry without regulation surpasses the regulated one occurs earlier. For example, if the static efficiency gain is only 5 percent, then a 1-point differential in growth will produce a crossover in five years. If the static efficiency gain is 1 percent, a 1-point differential in growth will produce a crossover in only one year.[34] If the static efficiency gain is 1 percent, a 2-point differential in growth will cause the high growth industry to outperform the other by 46 percent at a point twenty years out.

These figures demonstrate that the choice of a regulatory framework that recognizes the complex and dynamic nature of our economy—as opposed to one that is based on "static," intrusive regulatory policy—can be crucial. The power of compounding growth rates will probably assume even greater importance in the Innovation Age. Recognizing and dealing with the changed political economy of innovation has now become the real challenge in reinventing government.

Chapter Two
Telecom Revolution

Digital Technologies

A primary force behind the development of the Innovation Age is the greater availability of information and the accelerating pace at which it can be shared. No one can doubt that telecom improvements have tremendously enhanced our ability to share information. Thus, we can view telecom as having effectively increased the number of beneficial interactions between holders of new ideas. An increase in interactions should subsequently produce cross-fertilization of an initial innovative idea in wholly unrelated fields. Further, as in the handshake example described in Chapter I, the addition of one more information holder to the already large number of "interactors" existing today produces a very large increase in the number of interactions overall. In acting as a catalyst for beneficial interactions, telecom improvements have become a major force behind the Innovation Age. And in light of the developments now on the horizon in this industry, that impact on innovation is likely to intensify.

It is hard to ignore the revolutionary changes taking place in telecom. The revolution springs from dramatic advances in microprocessors (the brains of a computer) and other digital technologies that are also driving the computer industry. When digitized, a telephone call, home banking transaction, or television picture involves millions of tiny packets of information (made up of "1"s and "0"s) that computers and fiber optic cables can transmit, decode, and reassemble with incredible speed. The result is a remarkable improvement in capacity and quality over transmission systems available from existing alternatives. The advances that have occurred in microprocessors during the last decade have been breathtaking. The amount of information a microprocessor chip can handle has increased with the number of circuits a chip can hold. So-called chip density has doubled every eighteen months, producing a hundredfold increase to date, and this trend is expected to continue for at least another ten years. Rather than slowing—as some experts had predicted—the growth rate of density is probably accelerating. The 8088 chip, introduced twelve years ago, had circuits 3 microns wide; today's Power PC chip has circuits just 0.65 microns wide, and IBM/Siemens is shipping a chip with circuits 0.35 microns wide. Some experts predict that in the future circuits will be laid down molecule by molecule.[1]

The profound improvements in the cost and performance of digital technologies have made possible better and cheaper switching, terminal, and transmission facilities.[2] Telephone switches have become sophisticated computers that are hundreds of times more efficient than the previous, mechanical switches. Thousands of miles of fiber optic cable have been laid to serve high-volume routes and customers. Fiber transmits digitized data at little cost and can be cheaply improved by replacing the electronic equipment used with it. As a result, the capacity of optical fiber networks has been increasing by 50 percent per year since 1980. Driven largely by digital technologies and competition, productivity in the telecommunications industry is now believed to be increasing by at least 5 percent per year.[3]

Advances in digital technologies promise to continue the telecom revolution. In the near term, local telephone, cable, and other companies are likely to continue "digitizing" their networks and a variety of other transmission technologies that might be used to reach residential customers.

In addition to fiber optic cable and coaxial cable (used prevalently in cable TV systems), parts of the digital network will be "wireless." At the same time, new digital technologies will greatly expand the amount of voice and data communication that can be transmitted over existing channels, freeing up spectrum for other purposes. Digital technologies will increase channel capacity five-to-tenfold. Existing mobile radio services will be replaced by secure, versatile, high-fidelity digital services. All of this is coming at a time when the FCC has released large amounts of the electromagnetic spectrum for a new mobile personal communications service called PCS. Without these new transmission technologies, PCS and the other new wireless services would not be economically viable.

The pervasive effect of digital technology on telecommunications can also be seen in recent developments in high definition television (HDTV). HDTV would give viewers movie-quality video and sound on big-screen TVs. It should be noted that HDTV was originally forecast as an analog, pre-digital technology that would be dominated by Japanese firms—the result of a Japanese government-business consortium. American firms, much to the chagrin of industrial-policy advocates in the U.S., chose to leapfrog the analog system and develop a digital HDTV system. Exploiting the revolution in microprocessors, the "Grand Alliance" of Zenith, AT&T, General Instrument Corp., Philips, David Sarnoff Research Center, Thomson, and MIT have finished testing a digital HDTV standard that outperforms the Japanese-developed analog system in picture quality, uses less spectrum, and can meet the growing demand for improved video resolution in computing applications.[4]

In the past, the main impact of these digital improvements on the telecom industry has occurred within the telephone companies and has been manifested to consumers indirectly in the form of lower prices. Increasingly, consumers will feel the impact of continuing reductions in transmission and switching costs directly in the form of new services and competition in local telecom markets.

Competitive Benefits

In addition to continuing advances in digital technologies, recently passed telecom legislation will open local telephone markets to competition, further propelling the telecom revolution. The telecom market has long been highly regulated, but the rise of competition and deregulatory measures such as the Telecommunications Act of 1996 are undermining the old order. Experience in this market as well as in other industries shows that introducing competition into previously regulated markets lowers costs, reduces subsidies, and spurs innovation. All these benefits can be expected from the development of competition in local telecom markets.

For sixty years regulation of the telecom industry has been based on the Communications Act of 1934 and on assumptions of monopoly and scarcity. The 1934 Act created the Federal Communications

Commission, giving it broad jurisdiction over the telephone and broadcast industries. Telephone companies engaged in providing interstate or foreign communications for hire were regulated as permanent monopolies and the reasonableness of their rates and profits was scrutinized by the FCC. In addition, radio and TV broadcasters and all users of what were thought to be highly scarce electromagnetic spectrum frequencies were licensed by the FCC. Broadcast licensees, in particular, were subjected to significant content regulation.[5]

In the process of deciding how frequencies were to be used and by whom, the FCC gave short shrift to competition and new services, protecting existing services instead. Billion-dollar examples of spectrum mismanagement are detailed in Chapter V. Costly regulatory delay also occurred outside the realm of spectrum allocation. Three illustrative cases of this delay involve the telephone equipment, long-distance telephone, and cable TV markets.

In 1956, the FCC backed AT&T's tariff prohibiting the sale of the Hush-A-Phone—a plastic attachment to the mouthpiece of a phone that afforded more privacy.[6] In fact, it wasn't until the late 1970s that equipment competition was permitted. Since then competition and innovation have burgeoned for sophisticated business phones such as PBXs, key systems, teleconferencing and fax equipment, as well as for residential cordless and cellular phones. In addition to foregone innovations, a decade-long delay of competition in the $16 billion per year equipment market undoubtedly cost consumers billions of dollars.[7]

FCC action also delayed competition in the long-distance market. In the 1970s, the FCC repeatedly prohibited MCI from trying to expand its narrow business offerings to provide general long-distance service.[8] Only after MCI twice prevailed in federal appeals court did the FCC permit it to compete fully.[9] As a result MCI, Sprint, and a host of other carriers began to compete vigorously and have now garnered close to half of the $65 billion long-distance market.[10]

The FCC also placed intrusive limits on what programming cable TV operators could and must carry on their systems.[11] These regulatory efforts were explicitly designed to handicap the ability of cable TV to provide viewers with additional programming (ironically, the FCC was regulating over-the-air broadcasting because it was supposed to be highly scarce). Eventually many of these restrictions were eliminated. Any delay of cable TV service was costly to viewers, however, because cable systems are an efficient means of distributing much highly-valued programming. Cable TV now generates $23 billion in annual revenues and provides Americans with over 100 new channels of programming.[12]

Besides delaying new services, the regulatory system for telecom often distorted prices and business costs. The growing competition in telecom markets can be expected to address these distortions, as has been demonstrated by experience in other deregulated markets. In *The Economics of Competition in the Transportation Industries*, economist John Meyer and his colleagues trace the development of regulatory techniques that load costs disproportionately on to disfavored groups.[13] For example, in the case of railroads, "value of service" regulation shifted costs from bulk shippers to shippers of high-value items. Because transportation costs were less important overall to high-value shippers, that group was less likely to complain to regulators. Further, bulk shippers included politically powerful agricultural and mineral interests. This approach—i.e., benefiting preferred users—spread to the regulation of other industries, often favoring low-volume, infrequent users. In trucking, favorable regulatory treatment was given to small-lot and occasional shippers relative to large-volume, full-truckload shippers. Similarly, under airline regulation, "short-haul airline passengers (traveling under two hundred miles), particularly those flying between low-density city `pairs,' had their way subsidized by long-haul passengers flying nonstop between widely separated large urban centers."[14]

To maintain these artificial rates, market entry and competition in these industries had to be constrained. Thus, regulation reduced efficiency directly by distorting prices, and indirectly by impeding innovation and normal marketplace incentives that operate to lower costs. The extent of these losses, as discussed below, only became evident when the railroad, trucking, and airline industries were substantially deregulated.

In a similar fashion, telecom regulation has departed significantly from cost-based pricing to achieve particular social goals. Most important are universal service policies, the general purpose of which is to make telephone service affordable for all Americans. In practice, universal service has been used to justify several broad gauge federal and state regulatory policies that have acted to keep residential telephone rates, especially in rural areas, lower than they otherwise would have been. Subsidies are given to local telephone companies with higher than average costs rather than to individual households that might need assistance to pay for telephone access.

To generate the revenues needed to accomplish this goal, prices charged to long-distance, business, and urban customers have been higher than would otherwise prevail under competition. Long-distance companies give roughly $.60 per month per subscriber to the Universal Service Fund (USF) to subsidize higher service costs, generally incurred by rural telephone companies. Under this program, local companies can start requesting a subsidy when their costs exceed 115 percent of the national average. Besides discouraging efficient operation, the program also seems to cover too many companies—more than half of all local telephone companies, about 675, now qualify. Currently, the USF and related low income assistance programs redistribute roughly $800 million a year.[15]

Of much greater consequence are the subsidies built into the access charges that interstate, and some intrastate, long-distance callers indirectly pay local telephone companies for originating and terminating their calls. These subsidies are about 10 times as costly as those just discussed. Additional subsidies are built into the higher rates that many local telephone companies charge business customers for the same local service provided to residential customers.

According to the FCC, U.S. consumers spent an average of $58 per month on telephone in 1993, $18.82 for basic service and $39 for long-distance and other services—an increase of 5.4 percent from 1992.[16] Of that $39 per month, according to long-distance telephone companies, roughly $17.50 went to local telephone companies for the use of their facilities to originate and complete most long-distance calls. Meanwhile, long-distance and regional Bell companies report that from $12 to $14 of this average monthly long-distance bill went to subsidize local rates.[17] Currently, the local rates that many residential customers are paying are artificially low. According to NYNEX, in New York City the monthly cost per line is $22, while customers pay only $11 per month for basic service. US West Inc. says a phone line costs it $22 per month to maintain, on average, while customers pay a base rate of only $14 per month.[18]

It is likely that the customers paying the high margin rates will be targeted by the new competitors that are entering the local market—customers who could move much of their telephone business off the local network. The revenues at risk appear to be enormous. Economists Calvin Monson and Jeffrey Rohlfs estimate that

> the contribution [to overhead] for these [competitive] services is at least $18.3 billion, but not more than $21.1 billion. On average, it is over $12 per month for every access line in the United States.[19]

Alternative local telephone service providers question these estimates of the size of the subsidy. They contend that the precise nature of the subsidies is unknown because the revenues flow directly to companies. They also argue that local telephone companies are overstating the subsidy amount that is required since competition and new technology would reduce the cost of serving customers in high cost areas. They contend that, if customers used the most efficient technology available today, the cost of providing service in high cost areas would be as low as $7 billion.[20]

Regardless of which side is right, it is likely that competition will require changes in the scope and funding for universal service. As competition intensifies, the prices charged to the high margin customers will come under enormous pressure. Moreover, competition will inevitably spur significant rate restructuring, cost reductions and innovation.[21] As we will see in Chapter V, regulators will have an enormous impact on the extent and timing of these competitive benefits. In any case, substantial improvement in U.S. telecom performance is likely. Moving telecom prices into better alignment with costs and increased innovation will stimulate long-distance usage as the prices for those services fall, thereby increasing the relative significance of the services. Similarly, competition in the telecom industry will force local telephone companies to trim costs to match the lower costs of the new competitors. The new entrants will, on average, be using newer technology and have leaner labor arrangements.

The cost cutting efforts of major telecom companies are already well under way. From 1984 through 1995, the seven regional Bell companies and GTE reduced their collective payrolls by 150,000 jobs. In January of 1996 AT&T announced that it was cutting 40,000 jobs, and industry analysts have estimated that the Bell operating companies may have to reduce by an additional 50,000 jobs.[22] Competition will also stimulate the development of innovative wireless technologies that can serve sparsely populated areas more cheaply. For example, in rural areas competitors could use satellites and wireless facilities, lowering the cost of serving a subscriber by as much as $20 per month.[23] Similarly, a cellular company representative testified before a board of federal and state regulators that the average rural telephone company charges $1,800 per new installed line and that cellular carriers could cut that price by $700.[24]

Experience with competition in various transportation markets supports this point. Trucking deregulation, for example, reduced prices, especially for high-volume shippers of high-value products. Because the regulation of the trucking and railroad industries had increased shipping prices, it had become profitable for many shippers to avoid paying these rates by becoming "private carriers," that is, by owning and operating their own fleet of trucks. With deregulation of the trucking industry in 1980, shippers began to rely less on private carriage—presumably because competition had forced the rates paid by high-volume shippers to decline substantially.[25] Likewise, since deregulation the airline industry has been characterized by highly rivalrous pricing and other forms of competitive behavior. Half-price summer fares, sharp discounts for weekend trips, generous frequent-flyer programs, and free tickets for companions are now familiar competitive practices in the airline industry. From 1977 to 1992 the price per passenger-mile, adjusted for inflation, declined 35 percent.[26]

Competition and deregulation also forced firms operating in deregulated industries to reduce their costs. The U.S. domestic airline industry, in the years since deregulation, provides graphic evidence of this effect. It has become more productive than its foreign counterparts; indeed, it has grown seven times faster than the

U.S. population.[27] And airline competition continues to bring about cost reductions because full-service airlines are having to match fast-growing low-cost operators.[28]

It is evident that regulatory restructuring that has occurred in other industries and can be expected in telecom is remarkably powerful. Together with technological advance, this restructuring could cause a revolution in telecom over the next ten years that is no less important than the one that began in the computer industry a decade ago. Indeed, experience over the last decade with computers can be a guide to the future development and consequences of the digital network. Chapter III discusses how computer developments provide a useful paradigm for the ongoing telecom revolution. As we will see, the two industries are technically and economically related, and the insights of Hayek and Schumpeter apply to both.

Chapter Three
Computer Paradigm

Unanticipated Revolution

The revolutionary changes in microprocessors and other digital technologies that have driven the computer industry are the same forces that are now driving telecom. As discussed earlier, the rapid technical improvements in the performance of the microprocessor "brain" of the computer have greatly improved the underlying facilities used to provide telecom services. In addition to these technical links, other connections can be seen. Falling microprocessor prices have reduced the costs of generating information, leading to greater information production. At the same time, falling telecom prices have increased the demand for interconnected computing. In other words, the computer and telecommunications industries also have economic links.

The interconnectedness of the computer and telecom industries is such that the two industries can generally be considered separate segments of the same digital network. Unlike telecom, however, the computer industry has been largely free from regulation in the form of price and profit controls, government standard-setting, and mandated subsidies. Therefore, experience in the computer industry can provide useful insights as to what can be expected and what might prove possible as competition continues to grow in telecom markets.

The computer industry exemplifies the new relevance of Hayek's spontaneous order and Schumpeter's creative destruction. The vast changes in the industry have involved complex developments, most of them largely unanticipated. Notwithstanding this complexity, the market has achieved a powerful new spontaneous and creative order that has benefited consumers enormously. This section traces the uncertain and explosive changes in the computer market over the last decade as a result of the development of the personal computer. The section on technical uniformity examines how the computer industry, without the help of government, has efficiently set technical standards. The section on reasonable prices shows how the rapid pace of change in the computer industry has served to counteract the dominant status achieved by particular companies in some segments of the market.

Growing Uncertainty

The first thing to note about recent developments in the computer industry is that they were largely unforeseen. In retrospect, the principal reasons for this myopia were the revolutionary nature of quality and price changes in microprocessors, and the unanticipated effects that these improvements had on demand and the competitive balance in related industries.

The transformation of the personal computer during the last fifteen years has been nothing short of miraculous. The original IBM PC, introduced in 1981, boasted 64K of random access memory (RAM), two 160K disk drives, a monochrome monitor, and an 80 character-per-second printer; it sold for $4,385. Since moving from iron loops to magnetic media, computer memory has increased by 30 percent per year; in 1991 and 1992, it grew by 60 percent per year.[1] Current personal computers typically include at least 8,000K (8 megabits) of RAM and an internal disc drive of more than 500,000K (500 megabits) and cost less than $2,000 with a color monitor. Today's typical IBM-compatible personal computer, using a Pentium chip, is rated at 100 million instructions per second (MIPS), the same processing speed as a typical mainframe of an earlier generation; and the chip sells for less than $1,000.[2]

These explosive improvements in the price and performance of personal computers have created a dramatic chain reaction that has spurred personal computer sales and fostered investment in new software applications that could better exploit the greater speed and memory. The availability of these new software applications, in turn, has led to further computer sales, and so on. Thus, the interactions between technologies and across related industries have accelerated change.

The sheer size of these changes has created additional unpredictability. According to economic theory, the sensitivity—or "elasticity"—of demand to price changes increases over time, as consumers and businesses have more time to adjust. Initially, a new technology simply substitutes for an older technology. Only later is the technology used for entirely new purposes that require institutional changes to maximize their impact.[3] Therefore, the larger and more sustained a price change is, the greater and more unanticipated the uses of the product will be.

The price changes in computing have been large and sustained and the ripple effects powerful and unexpected. The flourishing of the once-lowly personal computer has undermined the mainframe computer market, devalued billions of dollars in market capitalization, caused company restructurings, evoked astonishing growth in upstart firms, and left government antitrust and industrial planning actions in a wake of silicon dust.

PC Juggernaut

Although it is difficult to imagine an office today without personal computers, the IBM PC was in fact introduced only fifteen years ago. At the time, many experts believed that the PC was destined to fill a narrow niche. On the day the IBM PC was released in 1981, company officials predicted that they would sell 250,000 units in five years. By 1991, they had shipped more than forty times that number.[4]

IBM certainly did not expect to see PCs erode the market share and profitability of its venerable mainframe computers, but the fall of the "big iron" machines was rapid and spectacular. The revenue of IBM's mainframe unit fell from roughly $9 billion in 1990 to an estimated $4.5 billion in 1994. IBM has now reduced its mainframe manufacturing capacity from twelve plants to four and has slashed 3,000 jobs in mainframe manufacturing.[5] Its mainframe prices are estimated to have fallen 30 percent from 1992 to 1993. Though still one of the world's largest manufacturers of computers and currently on a modest upswing, in the space of five years after 1987, IBM lost two-thirds of its market value—more than $70 billion.[6]

"Big Blue" is not the only mainframe maker to have felt the squeeze. The Japanese firms Fujitsu, NEC, and Hitachi have also suffered.[7] Digital Equipment Company (DEC), the manufacturer of the widely used VAX minicomputers, experienced a devastating $2.5 billion loss in 1992. And DEC's hard times have continued: it lost $135 million in the fourth quarter of its 1994 fiscal year before taking a special charge of $850 million to cover the layoff of 20,000 workers in its 1995 fiscal year. During the fourth quarter of FY 1994, DEC fired 7,000 workers.[8] DEC's CEO, Robert Palmer, summed up the mainframe business's troubles: "We all have business models that were built on very high hardware margins and those margins are no longer sustainable, so neither are the business models. And we all have that headache."[9]

Computer makers were not the only ones caught flatfooted by the downfall of the mainframe. In 1981 the Department of Justice was still pressing an antitrust case against IBM while market forces were about to lay waste to the company. Similarly, the Japanese, once thought destined to take over the computer industry, were caught off guard. Rather than designing new chips for PCs, the Japanese focused on building their own supercomputers as a means of taking over the high capacity segment of the market. But while the strides the Japanese have made in supercomputing are impressive, they have occurred at a time when the demand for supercomputers has substantially declined in favor of other technologies.[10]

Indeed, the parties destined to become players in the PC revolution were unknown when the PC was introduced, and the experts' prediction of a much-ballyhooed computer face-off between IBM and AT&T never materialized. Innovative companies that did not exist at the beginning of the revolution rose rapidly. Few people had ever heard of a small company named Microsoft. Nor had they heard of Intel, Novell, Compaq, Dell, or Netscape. Collectively, in early 1996 these companies had a market capitalization totaling more than $120 billion.[11]

Technical Uniformity

Computers around the globe interoperate because people have established standards that allow them to exchange information with one another. There are many examples in the history of this industry where the market, not the government, has set logical and efficient standards. In some instances, consumers and businesses have converged on one or two proprietary standards. In others, computer manufacturers have coalesced around uniform nonproprietary approaches.

While permitting technical convergence, this unregulated course has fostered innovation. The lack of mandatory standards has made it easier to introduce new technical approaches. It also has encouraged innovation by holding out the prospect of developing a lucrative new proprietary standard. This section considers how consumers enjoy the benefits of innovation and technical uniformity in three critical aspects of the personal computer: (1) architecture or system design, (2) interoperability, and (3) interconnectivity.

Open Architecture

The term *architecture* refers to the design of a computer system. In the early days of computing, system architecture was proprietary—that is, the property of a particular firm—and was protected by patents, copyrights, or other legal protections. Proprietary architecture may not be copied without the express permission of the firm that owns the property rights. Open architecture, on the other hand, is a system in which no firm owns the right to the design. The design is in the public domain, and anyone may copy it legally.[12] Despite the industry's early preoccupation with proprietary design, the history of computer architecture and operating systems reveals the existence of powerful incentives to produce technical convergence of design.

The first computer—the Electronic Numerical Integrator and Computer, or ENIAC—was developed by Remington Rand Corporation during World War II. Obviously, it was a proprietary system, because it was the only computer in existence. Soon, however, firms rushed to build electronic computers, and many were eventually successful. Remington Rand was joined by International Business Machines, Control Data Corporation, Digital Equipment Corporation, Texas Instruments, Unisys, Sperry, Wang, and countless others.[13] Each architectural design was proprietary and essentially operated as a closed system. A business typically would purchase a mainframe, software, and peripherals such as modems and printers from the same company.

Most mainframes, supercomputers, and workstations as non-open systems maintain the element of proprietary design. One *operating system*—i.e., the software that runs the computer—for these machines is UNIX, designed by AT&T's Bell Labs. UNIX became popular quickly because it is an open system, giving programmers at disparate R&D facilities greater flexibility. UNIX became the dominant operating system on most research machines, and the mainframe manufacturers developed proprietary versions of UNIX to optimize their individual hardware.

The philosophy behind IBM's first personal computer was fundamentally different. The PC architecture was designed as an open system. The open systems design was intended to encourage third-party vendors to get into the game of making expansion cards, peripheral equipment, and software. In the early 1980s, Microsoft and some subcontractors developed a UNIX-like operating system for IBM's personal computers; it was called the Disk Operating System (DOS).[14] This open architecture has spawned hundreds of software and hardware companies, and in large measure made DOS-compatible PCs the marketplace standard instead of the leading alternative system from Apple.

From the beginning, Apple's Macintosh line of PCs enjoyed a reputation for being more "user-friendly" than DOS-based machines, and once it was evident that Apple was in the market to stay, software development firms began to write software to accommodate Apple's operating system. Apple's system, however, was not open—Apple retained the exclusive right to manufacture and sell its computers. With no third-party "clone" makers pushing prices down, Apple's PCs were higher-priced than comparable DOS-compatible machines.

This price advantage for IBM compatibles triggered a consumer-driven development cycle that gave DOS-based machines a decisive advantage. The price advantage maintained DOS's edge in market share, and its greater market share led software developers to create products first for DOS-based PCs. This advantage was reinforced by consumers' understandable preference for PCs that could use the widest range of available software. Further, when the industry standard later changed from DOS to the Windows system, which runs "on top of" DOS, Apple lost most of its advantage in user friendliness.[15]

Recognizing that DOS was the de facto industry standard, Intel and most other chip makers focused their energies on improving microprocessors that could run DOS rather than the Apple operating system, a decision that further cemented the industry standard. By 1995, the market share for PCs using DOS-compatible (X86) microprocessors had risen to approximately 92 percent, up from 89 percent in 1993. Apple PCs, including those using a power PC microprocessor, now represent less than 10 percent of the market.[16]

Apple's failure to license its operating system to other manufacturers in the late 1980s was a strategic error. While Apple waited, Microsoft captured most of the business market with its Windows system, which some have argued was an inferior technology at the time. Apple's failure to license greatly restricted its ability to penetrate the market, while the open PC architecture on which Windows runs penetrated the market deeply. Apple is now attempting to battle back by licensing clones for its operating system, the MacOS.[17] Several companies now manufacture clones of the Macintosh and Power Macintosh lines of computers. Interestingly, both the clones and the Apple machines can be configured to run DOS, Windows, or Windows 95 in addition to the MacOS by adding hardware or software to the machine. This battle, however, will not be easy for Apple to win.[18]

The need for cooperation in setting standards is now well understood by computer hardware and software companies. Ray Noorda of Novell calls such behavior "coopetition." The idea is that computer firms that compete vigorously in the retail market can and should cooperate on standards. For example, IBM, Hewlett-Packard, and Sun Microsystems differentiate their workstations while preserving sufficient uniformity for UNIX programmers and users. This cooperation is essential because UNIX competes with the highly successful DOS/Windows software.[19]

Interoperability Standards

A concept similar to open architecture is interoperability. Interoperability refers to how computer components (such as the hard disk drive, keyboard, and monitor) interact with the microprocessor.

Whereas an open system allows other manufacturers to replicate components easily, an interoperable system allows parts made by different manufacturers to be interchanged one with the other. True interoperability allows consumers to purchase and install components and peripherals from a variety of vendors without worrying whether the various parts will work together.

In the mid-eighties, IBM tried to gain market share by creating a new data bus architecture, the Micro Channel Architecture (MCA). The data bus is the link between a microprocessor and the other components of a system. The original data bus was known as the ISA, or Industry Standard Architecture. MCA was a technologically superior design; it was wider (32 bits versus ISA's 16) and thus could carry more data. MCA, however, was not interoperable—MCA machines would not permit use of an ISA expansion card for adding a peripheral device to a computer.

MCA was an attempt by IBM to reestablish control over the PC industry. The industry response to MCA was a concerted effort to compete by several IBM-clone manufacturers. This consortium created a similar 32-bit bus known as the Extended Industry Standard Architecture (EISA). Unlike MCA, EISA is interoperable. Consequently, MCA architecture was beaten back, with one analyst claiming that "the ISA (and to a lesser extent EISA) standard is perhaps stronger today than in its early years. IBM continues to sell MCA machines almost exclusively, but has lost market share in doing so."[20] IBM's experience with MCA provides a powerful lesson. Even though MCA had a superior design, IBM underestimated its competitors' ability to respond. The response was stronger competition and an architecture that was interoperable with other systems and previous designs.

Interconnectivity Standards

Networks allow computers and their users to connect with one another. Currently, computers interconnect by way of analog or digital networks or some combination of the two. For instance, the backbone of the Internet is a confluence of fiber optic cables, wire, and satellite transmissions that allow data to be transmitted globally. Many major universities, research centers, and government agencies are directly connected to these cables. Other computer users connect to this network over traditional phone lines using modems. (A modem converts digital signals from a computer into analog signals that are transmitted across the phone lines. A modem at the other end of the connection reverses the process.)[21]

Computer networks with disparate designs can communicate with one another if there are some standard protocols on how data is transmitted. For example, Hudson Institute's Indianapolis and Washington offices, 500 miles apart, use different computer networks, Ethernet and Token Ring, but users on one network can send electronic mail and computer files to users on the other network without making any special efforts. In fact, the process is so seamless that most of the users are unaware that there are two networks, let alone which one they are on.

These seamless connections become critical as rival firms build their own networks using different materials and architecture. (The ability to use different materials—fiber optics, coaxial cable, copper wires, wireless systems—allows firms to produce systems at the least cost and thus highest productivity.) Subscribers to one network—for example, the local phone company—will demand the ability to communicate with users on another network, such as one provided by the local cable company. Users are also likely to want to be able to communicate with users on networks located across the state, country, or globe.

As a result, insular networks will not survive over the long haul. They will be overwhelmed by competitors that offer connections to other users, services, and databases around the world. Users on other networks will not be able to utilize the services offered by an insular network, which acts as an automatic brake on the market penetration of those services. As a rule, firms and architectures that do not penetrate the market deeply tend to be overrun by firms that do.[22]

This interconnectivity is what has made the Internet, a global network of thousands of computer networks, such a huge success. The system's increased access has broadened the horizons of millions of computer users. The computer labs on college campuses are no longer the domain of computer scientists. Students who know how to use e-mail are in the mainstream. E-mail has spread from the business world into elementary schools and homes, with thousands of new private users logging on each week.

Reasonable Prices

The pace of technological change in the computer industry, as we have seen in the previous discussion, has been extraordinarily rapid; at the same time, it has generated enormous consumer benefits and reordered the computer industry landscape. Generally, the cost and performance of the industry have been outstanding, although at various points innovative companies have achieved dominant status in one or another segment of the market. The normal "monopoly" consequences of the dominant status of companies such as IBM, Lotus, Intel, and Microsoft have been counteracted, however, by the power of innovation in a fast-changing digital world.

As prices continue to fall and hardware and software capabilities improve, the PC market continues to expand, undercutting the advantage existing firms normally derive from having a large embedded base of customers. A state of the art machine now can become obsolete in a few years or less. Consequently, there has been—and will probably continue to be—a constant stream of new customers entering the market for the first time and old customers looking to update their antiquated two-year-old PCs and software. Thus, a dominant firm knows that it must prove itself in the marketplace each day and that its current financial success only ensures that more companies will spend more money and hours attempting to topple it.

A dominant firm can lose its "King of the Hill" status in two ways. First, if it does not continually improve, it will lose market share and profits to low-cost imitators. For example, the ability of low-end PC manufacturers to make IBM clones fostered robust price competition in the PC market. Second, today's market leaders must worry that some established and well-financed competitor or possibly an upstart will produce a technical breakthrough that will displace them. This situation reflects that fact that gains from innovation are so powerful and beneficial to consumers that they outweigh the higher prices dominant firms can charge. Indeed, attempts to eliminate these high profits by regulating prices would almost certainly disserve consumers even if the regulations dampened the incentives for innovation only slightly.

Low-Cost Threat

Competition from the low-cost low-end section of the PC market revolutionized the entire computer manufacturing industry. As the price and performance of the personal computer improved, its market expanded dramatically. Increasingly, many of these new customers bought their PCs from IBM-clone manufacturers rather than the big-name ones. The low-end manufacturers operated on much thinner margins, offered lower prices, expanded their market shares, and improved the quality of their computers until their performance was virtually indistinguishable from that of computers priced 50 to 100 percent higher. The brand-name PC manufacturers were forced to cut costs dramatically. For example, starting in 1991 Compaq cut its work force by one-fifth and extracted even greater savings from materials.[23]

The experience of Intel, the leading chip maker for PCs, illustrates the threat low-cost competitors can pose. Intel has been both a low-cost supplier to the makers of clone PCs and a target of low-cost chip makers. It has encouraged competition between low-cost and name brand PC makers by promoting its microprocessors directly to consumers, thereby affording instant credibility to clone manufacturers who use Intel microprocessors. It also pays up to 50 percent of the cost of a manufacturer's advertisement if the ad includes the phrase "Intel Inside." To the annoyance of Compaq and others, these practices make it harder for brand name computer manufacturers to differentiate their products from low-end producers on the basis of quality.[24] But Intel's incentive is to foster competition in the PC market in order to keep prices and margins low so that it can sell more microprocessors.

Intel itself faces a low-end threat in the microprocessor market. The company currently dominates,[25] but Compaq, Intel's largest customer, for some time resisted Intel's efforts to sell Pentium microprocessor chips. During this period Compaq's Impresario machines used 486 microprocessors purchased from Advanced Micro Devices. Compaq believed that the 486 chip was more than adequate for the then-existing software applications. Moreover, because Intel's edge over its competitors was far smaller on 486 chips than on Pentium chips, Compaq got a much better price on 486 chips. Compaq has also invested in or assisted two other Intel rivals, Cyrix and NexGen.[26] Rival manufacturers of chips for DOS-based machines, though tiny compared to Intel, continue to turn out quality chips, although until April 1996 none was able to do so without licensing Intel's technology.

To counter this competitive threat from the low end, Intel cut its prices on Pentiums and actively promoted the development of new software applications that required the power of the chip. Rather than sitting on fat gross margins—up to 80 percent on high-end Pentium chips—Intel slashed prices on its top-of-the-line chips, in some cases by as much as 40 percent in only four months. The top-of-the-line Pentium, when only in its first year, sold for 40 percent less than the 486 chip did in its second year.[27]

High-Performance Threat

The history of computers is filled with examples of market leaders who lost their dominant position or suffered major losses because innovative new products or services undercut them. Innovation in microprocessor design has caused major disturbances in the mainframe market as well as in the brand-name PC market. For example, in the late 1970s and early 1980s, new minicomputers (smaller-scale mainframes) began to consume part of the mainframe market, but workstations (one step down in speed and capacity from the mini's) began to bite into the minicomputer market by the late 1980s and early 1990s. As the power of microprocessors grew and the sophistication of personal computer networks increased, PCs began to supplant workstations. The resulting competition forced big cuts in costs and prices as the "umbrella" pricing of the market leaders collapsed, first in mainframes and minicomputers, then in high-end brand-name personal computers.[28]

This experience shows that, even without a threat at the low end by clone-makers, companies such as Intel cannot rest on their technical laurels. The threat of innovation at the high end by would-be competitors is a constant spur. Even Intel's dominant position in microprocessors is not safe. Intel has consistently cannibalized the architecture of its chips for new designs that have pushed the envelope of technological prowess. There is every indication that both Intel and its competitors will continue to do so in the future. As an example, Intel has announced a joint venture with Hewlett Packard to design microprocessors for personal computers that are four technological generations beyond today's computers. These new machines will be rolled out near the end of the decade.[29]

Meanwhile, Intel's competitors are pursuing their own market strategies. Advanced Micro Devices (AMD) has now begun shipping chips that are compatible with Intel's Pentium. The company developed the chips from the ground up and is selling them for as much as 50 percent below Intel's prices.[30] The company's new chip is called K5 (the "K" stands for kryptonite, the nemesis of Superman). It does not rely on Intel's technology and runs at a clock rate between 75 and 90 MHz.[31] To compete at the high end with Intel's Pentium Proline, AMD bought NexGen Inc.,[32] which is expected to bring out a 686 chip in early 1997.[33] Another competitor, Cyrix, plans to launch a new line of PCs to give consumers first-hand experience with its own microprocessor chips. Some studies show the chips outperform the Intel originals.[34] Also, AT&T has announced that it will make a chip that allows personal computers to be used as video phones. A consortium of high-profile computer companies such as IBM, Apple, and Sun (but not Intel) has agreed to build equipment compatible with the chip.[35]

Innovation has created the same rivalrous environment for computer software. Major software manufacturers that once held dominant positions in specific applications have had vast chunks of market share wrested from them by innovative competitors. For instance, WordPerfect once dominated the word processing market, but a significant portion of its clientele was stolen away by packages (such as Microsoft's Word for Windows) that were easier to use and featured "what you see is what you get" displays. Similarly, Lotus Corp. saw its 1-2-3 spreadsheet lose market share to competitors whose packages were thought to offer superior features, such as Quattro Pro and Excel.

Even Microsoft's DOS/Windows is not immune to attack. Microsoft has more than 70 percent of the world market in operating systems for personal computers and a 50 percent world market share in PC applications.[36] To date, it appears to have fended off all challenges, including one from IBM's OS/2 operating system. However, IBM and Apple have formed an alliance to develop hardware built around the PowerPC chip that would run both Apple and IBM software, such as Apple's Power Macintosh Operating System and IBM's OS/2.[37] Computers based on this chip would not require users to choose between Apple or DOS-based software when buying a PC, because they could run both. As a result, Apple machines would be able to compete more effectively against Intel-based machines, especially in the corporate market. Although this effort is apparently struggling,[38] if the new IBM PowerPC were to become popular, Microsoft's dominant operating system would be more easily challenged. This is because software developers for Apple and OS/2 applications would be guaranteed that a universe of machines broader than either the IBM PC or Apple alone could use their products.

Of late, two new lines of attack on Microsoft have arisen: first, from IBM, which purchased Lotus (principally to acquire Lotus Notes) and second, from new uses of the Internet. Lotus Notes is an extremely popular piece of network user software. Part of its attraction to IBM comes from the fact that Lotus Notes has become a new interface between the user and his PC. At the moment, however, both Lotus Notes and Microsoft are facing competition from Netscape and other makers of "Web browser" software. Netscape allows users to put together company networks using the Internet or "intranets" at a fraction of the cost of using Lotus Notes.[39] Now AT&T and MCI have formed alliances with various computer and Internet browser companies in order to compete in providing these intranet networks which link employees within companies.[40]

It is also thought that Internet-facilitating companies such as Netscape could pose a significant long-term threat to Microsoft's dominance of key technical standards. The Web can serve as a new platform for software development because Sun Microsystem's programming language, Java, allows users to download software from several operating systems in addition to Microsoft's. Thus, users can order up customized software from numerous central data bases on an as-needed basis. Hundreds of Java applications or "applets" already exist and the rise of powerful computer networks makes their use more feasible. Some observers even predict that many consumers may choose to buy $500 PCs with less memory and rely on applets.[41] While this scenario may be improbable, the popularity of Java has already led Microsoft to license Java from Sun. Microsoft's 1997 version of its Internet Explorer browser will be offered as a free add-on for users of Windows 95 and Windows NT. As a result, Microsoft customers will be able to read pages of Web information created with Java. This development represents a dramatic departure from Microsoft's reliance on its own technical standards.[42]

The upshot is that although some firms may dominate a particular segment of the computer market and temporarily may be able to charge prices above those that would exist in a perfectly competitive market, the rapid rate of digital innovation largely invalidates traditional monopoly analysis. Above-normal profits for a market leader motivate rival firms to attempt to displace it. The dominant firm, for its part, must continually look over its shoulder and innovate to stave off the advances of rival firms. This process creates an "arms race" in innovation, because innovation is the key to attaining or maintaining dominant status in the digital arena. Even over the short haul, the threat of competitive innovation generates price and performance improvements that are far more significant than the consumer welfare (i.e., deadweight) losses attributable to prices being above the leading firm's current costs.

The computer industry has established itself as the paradigmatic industry of the Innovation Age. Its evolution has involved an enormously complex series of parallel and sequential technological and economic developments. No one individual or firm foresaw even a fraction of the consequences. Largely unregulated market forces, however, spontaneously achieved a hugely beneficial order of lower prices, greater productivity, and the best technical standards available.

If ever an industry exhibited "creative destruction," it has been the computer industry. Consumers have benefited enormously even though at any one time a few large firms have dominated the industry. For example, as Microsoft and Intel have gained market share, the price/performance ratio of their products has improved dramatically.[43] Consequently, where personal computers were once the province of large corporations, they are now more readily affordable for individuals. If today's dominant firms act as if they have a permanent monopoly, they run a real risk of becoming the target of the market's next round of creative destruction.

Chapter Four
Telecom Ramifications

Hayekian Flux

The creation of Hayek's "spontaneous order" is an unpredictable process. Who would have predicted that today the third largest commercial banking center in the U.S. after New York and San Francisco would be Charlotte, North Carolina; the third largest film and TV production center after Los Angeles and New York would be Orlando, Florida; and the second largest concentration of computer software programmers after the Silicon Valley would be in Provo, Utah?[1] Today's pace of innovation is blinding both in its speed and its effect on our ability to see into the future. We cannot be sure what is going to happen in one narrow field of technical endeavor, let alone begin to see the full societal consequences of the wide range of technological changes now under way.

The closer one is to the process, the harder it is to see where it will lead or place it in proper context. This is especially true in periods of great turmoil when society is confronting a wide variety of changes. Citing Juan Huizenga's *The Waning of the Middle Ages*, House Speaker Newt Gingrich illustrates this point using the Renaissance. What seems to us "a brilliant, exciting period of innovation seemed to its contemporaries a terrifying collapse of the existing order."[2] The changes occurring in today's uncertain world may well be as difficult for us to place in perspective as those changes at the end of the Middle Ages were for the people living through them.

Nowhere in our modern economies is the state of flux greater than in the area of digital telecommunications. Excitement over the potential of the "information superhighway" has reached a fever pitch in the past few years. Numerous groups have touted whole new categories of services, promising the ability to search the Library of Congress from the comfort of home, order up movies "on demand," and play video games with partners from across the country. Still, all the hype about new telecommunications services tends to mask the uncertainty surrounding these developments. Even the phrase "information superhighway" is highly misleading, conjuring as it does images of an essentially unchanging, turn-of-the-century technology and monolithic infrastructure, the path and capacity of which can be centrally planned and laid out by government engineers as were the interstate highways.

Nothing could be further removed from the truth. As discussed earlier, the changes in telecom are a manifestation of the same advances in digital technologies that are driving the computer industry, not a twenty-first century version of the railroad or highway. The new digital services will likely be limited and shaped by the uncertainty, potential for explosive change, and other forces that characterize the computer industry. No one knows what kinds of electronic traffic, let alone how much, will actually travel through digital networks. Although we can predict that "winning" technologies and services will be determined by individual firms responding to individual consumers, precisely what services will be popular in the marketplace and at what price is impossible to forecast accurately.

Similarly, the pace at which competition will develop in telecom markets previously dominated by protected monopolies is largely unknown. Nonetheless, consumers are already beginning to face a choice of competing local voice and video companies, overturning many of the assumptions upon which regulators and telecom companies have long relied.

A wide range of technical and economic issues affecting the supply and demand for these new services, especially interactive services, is as yet unresolved. For example, will high capacity interactive video services be delivered to the home? Will users be able to rapidly send as well as receive video? Will the services be provided by telephone fiber, cable coax, DBS satellite, or on wireless facilities? How much, if anything, will the average consumer be willing to pay for information services such as online shopping or the ability to search remote libraries or read magazines electronically?

Two supply-side issues are important to the success of the new digital services: their relative cost and their technical feasibility. Today, the cost to the home of the new technologies, especially high capacity interactive video, is substantial. Meanwhile, good substitutes exist for many of the most intuitively attractive new services. At the present time, 85 percent of all households have a VCR. More than 60 percent of all households subscribe to cable TV, and over 90 percent have cable facilities available. In roughly one year, more than one-million homes have signed up for DBS satellite service, and the number may go as high as 12.5 million by 1998.[3] Existing services such as TV and radio outlets can achieve some interactivity by relying on existing telephone service.[4]

Online services are another available choice for interactivity. The major online service providers claimed eight million subscribers in 1995.[5] According to an extensive October 1995 Nielsen survey, the number of people using all commercial online services was twelve million, up from seven million since the beginning of that year.[6] The Internet, used for electronic mail, intercompany collaboration, information gathering, and direct marketing and advertising, has grown phenomenally. The same Nielsen survey found that thirty-seven million people in the U.S. and Canada have access to the Internet and of that group twenty-four million are Internet users. The latter number represents 11 percent of the North American population over the age of sixteen.[7] The burgeoning growth of the Internet may now be undermining online services as users find they can get what they want using a cheaper Internet access provider.[8] The new applications of Internet are legion.[9] The Internet can even be used to make long distance calls.

The cost of providing consumers with high capacity video services such as movies-on-demand and video e-mail has been estimated to be as high as $500-$700 per home.[10] The cost of a set-top controller box has been estimated to range from $190 to $400 depending on its features and sales volume.[11] The cost of the current experiments is even higher. A NYNEX video services executive states that that company's planned "video dial tone" (VDT) test networks will cost $1,500 to $2,000 per household, although the per home expense will come down dramatically as more equipment is deployed. Time Warner has estimated that it will cost $5 billion to install a fiber optic "backbone" into its cable systems around the country.[12] To put these cost estimates in perspective, even if video-on-demand services were to capture 50 percent of the video store rental market, it would take seventeen years for network builders to break even.[13]

Some companies are waiting on interactive video, while others are proceeding with bold wireline and wireless strategies. Michael Luftman of Time Warner Cable says, "Things are happening so fast right now that if you wait around to do all the classic testing and evaluation, you'll miss the boat. . . . We've already made the commitment to rebuild all of our operations into fully interactive networks."[14] Other companies have announced that they are modifying their interactive video plans. Bell Atlantic has chosen not to pursue its video dial tone plans using existing copper telephone lines. PacTel is scaling back its proposals to build interactive video networks throughout California and will rely on "wireless cable" frequencies to provide television service instead. PacTel had originally planned on reaching a total of 5 million homes with its interactive video network by the year 2000. Now it projects reaching three million homes with wireless cable by 1997, and five million homes by the end of 2000. NYNEX and US West have also scrapped aggressive construction plans. Ameritech is considering acquisition of traditional cable systems.[15]

Similarly, some cable companies initially justified the prospect of huge outlays for networks by pointing to the potential of an information superhighway that would let people order movies and clothes and play games, all via a souped-up cable system. But that technology has proved daunting to master, and the cable companies have turned their focus to the more immediate potential of the phone business.[16] As for the new services, Edward Horowitz, the CEO of Viacom Interactive Media, believes it makes more sense to "focus on small audience bases" and to make sure that there is validation of the business models for new interactive markets. He sees multiple mini-markets and viable businesses in video games, CD-ROM, and online services, with interactive television further down the road.[17]

A second source of supply-side problems hinges on technical feasibility. Important software obstacles need to be overcome to make interactive services feasible. Based on Viacom's experience in building a full service network testbed in Castro Valley, California, interactivity has taken longer to build than anyone imagined.[18]

Although virtually all the hardware exists, including video servers, trial delays have arisen out of the difficulty of getting all the pieces to work together.[19] These complex systems need to be integrated to enable software developers to proceed.

Another issue concerning technical feasibility is that user-friendly designs must be developed so that customers are not faced with fifty-button remote controls or twenty-five menu screens.[20] As Walt Disney TV President John Cooke has stated, "The central issue is whether programmers can devise an appealing way to meet the needs, expectations, and desires of consumers."[21] Consumer acceptance may also be held up if systems are not interoperable so that customers can readily change carriers.[22]

Indicative of the uncertainty over developing feasible interactive systems is the lively debate over whether the television set or the personal computer will be the source of consumer interactivity. A Canadian cable industry official has stated: "It's definitely going through the television, not the PC."[23] This view was recently backed up by a consumer survey conducted for Hewlett Packard. Even among the most technologically advanced consumers—those owning PCs and other high technology devices—45 percent said they would like to receive these services over their TV sets, compared with only 16 percent who chose the PC.[24]

The opposing view is vigorously argued by George Gilder who cites the growth in the number of computers connected to networks to support his optimistic views of the prospects for interactive services. Similarly, at the Convergence 95 interactive TV conference, Dean DeBiase (CEO of ImagiNation Network) remarked that "TV isn't optimum for many interactive TV applications because the PC is faster."[25] Recent developments regarding the production of cable modems that would operate one thousand times faster than today's telephone modems support this view.[26] By the end of 1996, however, the number of cable systems equipped to use cable modems both to upload and download will be only 20 percent.[27] Still other analysts see multiple solutions employing both computers and TV as likely.[28]

In addition to the supply-side projections noted earlier, the evidence on the demand side paints an uncertain and somewhat gloomy picture of the prospects for many of the new digital services. Given the sizable investment needed to provide interactive video, it will be essential that a substantial number of consumers be willing to pay for the extra services that digital services can provide, such as video-on-demand. Survey evidence on consumer attitudes, market experiments, and other analyses generally raise doubts about the near term prospects for services that provide or require significant consumer interactivity.

The results of consumer surveys for services that do not yet exist and with which consumers have little experience are suspect, many experts believe, and polls on the subject have even turned up contradictory conclusions. Often, people appear to respond in ways they think they should. "One of the lessons we've learned is that people don't always do what they say they're going to do," says Chet Thompson of Prodigy. "Market surveys told Prodigy that people wanted to do their grocery shopping by computer. They didn't."[29] A Hewlett Packard survey found that providing movies-on-demand through cable TV would be a highly popular service, while a Lou Harris and Associates poll found that this service would be among the least popular.[30]

The results of market tests are also mixed. Although the ability to pull up stock quotes has gained ready acceptance, and even though video-on-demand may eventually prove successful, the results to date are not encouraging.[31] On the other hand, the reliability of the test trials has also been questioned. For example, an experimental system used in Cerritos, California was not fully interactive, nor was it designed with an eye toward expansion or commercial viability.[32]

The Cerritos test, operated by GTE, combined a cable TV channel with a home telephone line; this system offered users such services as ordering movies and plane tickets, shopping, transferring stocks, and doing research. The trial does appear to support the contention that consumers are relatively satisfied with their VCRs. Thomas Greib, GTE's Director of Interactive Services, says he has learned several lessons from his company's experiment. Customers may want movies-on-demand, but to get that service they won't pay more than about $1 more than retail charges at the local video store. Also, customers prefer paying a monthly flat fee for the overall interactive service to being charged for use by the minute, and they prefer paying $10 per month rather than buying a set-top controller box for $250 and paying $5 per month.[33]

Some analysts point to the low rate at which consumers use VCRs to time-shift programming as evidence that there is little demand for that capability. Even for most-watched shows, less than 10 percent of viewing is time-shifted. It is reasonable to believe that if consumers were really interested in time-shifting or avoiding ads they would learn and use the necessary programming commands that would allow them to do so.[34] Similarly, in spite of the growth in online services and the Internet, it is not clear that there are very many people who want to look up massive amounts of information in databases.[35]

Finally, many consumers may lack the technical interest and time to become heavy users of these services. For example, in a 1994 survey 90 percent of the respondents who did not own computers (then roughly 67 percent) reported that they were very unlikely to change that status.[36] Also, people are now working considerably more hours per year, and have less time for leisure,[37] which can mean that they have less time for experimenting with these services at home. Sumner Redstone, Chairman of Viacom, has put these views bluntly:

> The last thing most of us want when we plop down in front of the TV is responsibility. So picking the ending to movies or even camera angles of sporting events won't pay for infrastructure needed to provide such services. . . . It will take a lot longer and cost a lot more to create the so-called information superhighway, with no certainty that the consumer will pick up the price tag . . . after 40 years the only quasi-interactivity on cable is home shopping. . . . Viacom's own test in Castro Valley, CA is adding proof to what is fast becoming undeniable . . . there's no killer application that will pay for info highway. . . . [E]ven full video-on-demand buy rates up to 4 times as high as pay per view aren't sufficient to finance superhighway infrastructure.[38]

Given the current high state of flux and uncertainty about the specific effects of the telecom revolution, including the uncertainty regarding the prospects of interactive video services, what, if anything, can we conclude about what is likely to happen? The next section considers the telecom market's most likely bets for Schumpeterian creative destruction.

Schumpeterian Bets

The ramifications of the telecom revolution are difficult to predict. As we have seen, there exist important unanswered questions about cost, user friendliness, and consumer demand. In addition, the public discussion has been misleading. The alluring potential capabilities of high capacity interactivity have focused attention on new services that are glitzy but not yet grounded in economic reality. Press coverage necessarily overemphasizes the investment plans of leading businesses, which all too often tend to be overly orthodox and myopic. Even the public discussion of technological developments, which are fascinating and sometimes quirky, often gets caught up in the day-to-day technical and financial ups-and-downs.

The most solid basis for predicting the ramifications of the telecom revolution is to begin with an economic effect of which we can be confident. The evidence of continuing advances in digital technologies and increasing competition among telecom providers discussed in Chapter II make certain price and performance improvements highly likely. It is a safe prediction that telecom services will become cheaper, especially over long distances, and more mobile. So-called broadband applications, such as video, that require substantial capacity will become much cheaper. Telecom services will also have much higher fidelity and become simpler to use. Although it is not clear which technical modalities will prevail—fiber, coax, cellular/PCS, wireless cable, or satellite—high capacity video will become available to most businesses and many homes on at least an incoming basis. Users will find it easier to reach specific audiences.

Predictions based on these enormous improvements in the cost and performance of telecom services will be both more reliable and fundamental. On a basic and macroeconomic level, the ongoing telecom revolution will heighten the ability of the entire marketplace to serve as a discovery process. This will be a decisive factor in the Innovation Age. On a microeconomic level, the ability to communicate more effectively and cheaply can be expected to produce specific impacts on related industries.

Discovery-Process Effect

As a result of these telecom improvements, the sheer number of telecom interactions will burgeon. The amount of information exchanged will grow even faster. And the possible interactions, as well as information capable of being exchanged cheaply, will grow exponentially. A whole lot of interacting will be going on.

This ability to communicate more effectively and cheaply will have two consequences. On the one hand, in many cases the transaction costs of voluntary exchange will fall. That is, improved telecommunications will often reduce the market friction that keeps a particular buyer and seller from finding each other. It will cost less to strike a mutually beneficial deal.

On the other hand, at the same time that it expands the scope of possible interactions, the improved ability to communicate will increase complexity and the potential benefit of experimentation. In dealing with "information overload," consumers will find it cheaper and more efficacious to rely on various intermediaries to help them sort through their expanded opportunities. Moreover, voluntary transactions do not always turn out as expected. Sometimes they turn out better; sometimes worse. The ability to communicate more effectively and cheaply will better enable consumers and producers to learn from each other's trials and errors.

Both of these effects will sharpen the market's ability to act as a "discovery process." Marketplace friction will be reduced. Simultaneously, the boundaries of the marketplace and the complexity of economic transactions will expand. As a result, the importance of decentralized decisionmaking will become greater. The existing (or any given) level of prosperity will be achieved more cheaply and quickly, because the cost of transactions will be reduced. The level of achievable prosperity will also become much greater because "transaction costs" will be reduced and can be spent elsewhere and because the reduction of transaction costs will expand the scope of beneficial interactions.

Most often the benefit will come from slightly more efficient transactions that will sharpen the marketplace's ability to move scarce resources to their highest and best use—thus forming Hayek's spontaneous order. The marketplace will more efficiently measure how consumers and producers value the myriad inputs and outputs in the economy. It will also strengthen the incentive and ability to act on these values. The cumulative impact of millions of such accretions in value will undoubtedly be enormous.

In rarer cases, which will become less rare in the Innovation Age, the gain will come from "the new consumers' goods, the new methods of production or transportation, the new markets, the new forms of industrial organization that capitalist enterprise creates"—thus giving us Schumpeter's creative destruction.[39] Where, when, and to what extent these benefits will manifest themselves we cannot predict with any certitude. Nonetheless, the surest Schumpeterian bet is that the telecom revolution will heighten the comparative efficacy of the market as a discovery process.

Economic Effects

In addition to an economy-wide discovery-process effect, the ability to communicate more efficiently and cheaply will also have more focused results. It will, for instance, affect the retailing and distribution of related goods and services, especially informational goods that can readily be reduced to 1's and 0's. The added capacity these telecom improvements will allow should cause significant changes in the entertainment and advertising markets, making it possible to overcome current limitations in the broadcast television market. The availability of more channels and subscription programming will give consumers more choices, one of which will be to avoid advertising. The nearly ubiquitous interconnection of microprocessors will open up entirely new avenues of information management. These and some of the other plausible economic effects on related industries from cheaper, better communications are outlined in this section.

Not-being there. It is likely that the reduction in transaction costs will change the retailing of goods in situations where consumers typically do not find it useful or desirable to examine firsthand the item they want to purchase. Among these might be situations where few terms of sale are important and where consumers view shopping in person as too time-consuming or downright unenjoyable. The potential market may be huge. The mail-order catalog business, which does $70 billion per year, could provide a base for online shopping. Interactivity might also make it feasible to expand this market. For example, periodic shopping for bulky, fungible commodities delivered to the home could be a low-cost way to perform an otherwise time-consuming and tedious household chore.[40]

Even in those shopping situations where there is little benefit in "being there," the success of online shopping will depend importantly on its ability to offer price savings by eliminating significant wholesale and retail costs. For example, Peapod, a small Illinois outfit serving 700 grocery buyers in affluent parts of Chicago and San Francisco, relies heavily on local supermarkets. It charges $30 per month on top of supermarket prices.[41] If interactivity and computer-guided logistics could give manufacturers of branded packaged goods a more direct pipeline to customers, such online shopping might offer price as well as convenience advantages. An eye-catching, easy-to-use video service might be a logical extension of today's warehouse club.

Information Goods. Where a good's value is chiefly in the form of information that can be digitized and transmitted in high fidelity, traditional distribution of physical goods can be largely eliminated. The potential savings in retailing these kinds of goods will be even greater than those described above. The greater the physical distribution costs are, relative to the total costs of producing a good or service, the greater will be the relative savings. Ticket reservations, stock transactions, and banking are three obvious examples of transactions that do not require physical distribution.

All these information transactions are now occurring on today's telephone network. Online services already provide a wealth of useful stock information as well as the opportunity to trade stock.[42] Online banking could be used to pay bills, check balances, transfer funds, and comparison shop for loans, mortgages, and credit cards.[43] One indication of the potential growth of such commercial transactions is Microsoft's willingness to pay a 50 percent premium for Intuit, which makes the popular personal finance software Quicken. Microsoft CEO Bill Gates justified this acquisition price by pointing out that people write more checks than letters.[44] Eventually, these services could be widely used by consumers and have a profound effect on their markets. Although improved telecom network capabilities may not be essential to their adoption, they would make all these services much more versatile, easier, and quicker to use.

Intellectual Property. Another possible source of Schumpeterian change is the market for intellectual property. Video and audio entertainment products are good examples. Whereas production entails substantial costs, physical distribution is not necessary. If the holders of these forms of intellectual property can find ways to protect their music and movies from appropriation by those who have not paid, a high capacity digital network will permit consumers to make high fidelity copies at substantial savings over the current system.[45] Substantial savings will result even if electronic distribution is limited to "wired" retail outlets. In short, by reducing retailing and distribution costs, the telecom revolution will increase the demand for intellectual property.

The production of news and political speech as a relatively low-cost form of information could also be greatly affected. George Gilder has noted that newspapers are particularly well positioned to fill this new demand. Newspapers already provide a kind of interactivity: readers can peruse news and advertising at their own pace. In the future, however, delivery via the computer screen—besides increasing the timeliness of the product—will increase the quantity of information that can be made available in the form of archives and other supporting material. Gilder says, "The ultimate reason that the newspapers will prevail in the Information Age is that they are better than anyone else at collecting, editing, filtering and presenting information, and they are allying with the computer juggernaut to do it."[46]

Narrowcasted Programming. The telecom revolution will also affect market processes by increasing both channel capacity and the number of programming distributors. The distributors could include independently owned TV stations, cable operators, wireless cable systems, DBS systems, and telephone companies that may provide much of their capacity to independent programmers. Increasing a distributor's channel capacity will make it feasible to provide more niche programming that can attract new viewers. In many local markets, increasing the number of programming distributors will also increase the availability of niche programming.

Currently, the existence of limited numbers of television stations in local markets leads to advertiser-supported programming that predominantly serves majority tastes. An advertiser-supported station maximizes ratings; it does not reflect viewers' intensity of preference. In this world of limited programming distributors, advertiser-supported media may all choose to cater to programming of interest to a majority of the total audience (although this programming may still be the second choice of everyone in the audience). For example, it might be more profitable to divide the situation-comedy market equally rather than capture the entire opera, outdoor sportsmen, or kids market. Professors Krattenmaker and Powe illustrate this point for a hypothetical world with only three TV stations and two types of programming. The first type of programming, say situation-comedy, attracts 75 percent of the total audience. The second type, opera programming, attracts 20 percent of the total audience. Assuming comparable advertising rates and production costs for both types of programming, it will be more profitable for the three stations to divide the audience for the more popular "sitcom" programming three ways than for any one station to get all of the audience for the less popular opera programming (75 percent/3 > 20 percent).[47]

The telecom revolution, besides improving picture and sound quality, will increase the number of independent programming distributors. This will make many more program choices feasible. Limited station numbers will no longer push advertiser-supported programming towards programming of interest to the majority of the audience. In the previous example, for instance, ten stations would make it profitable to carry the alternative type of programming (75 percent/10 << 20 percent). More and more consumers will be able to get their first choice in programming. The least-common-denominator programming market will be saturated. Program diversity will be limited only by the need to cover programming costs, not by the number of stations. Finally, the ability to attract viewers to least-common-denominator programming will be undercut by the availability of more advertiser-supported niche programming that is the first choice of viewers who previously watched the least-common-denominator fare.

Subscription Programming. The availability of subscription media can also make it profitable to cater to minority-viewing tastes. The development of cable TV systems has provided subscription mechanisms and greatly diversified programming content. There are currently more than a hundred cable networks. Also, pay per view channels allow relatively small groups of viewers to see "big events" if they are prepared to pay enough.

As a result of the telecom revolution, increasing amounts of programming will be available exclusively on a subscription basis or as an option. This movement will be driven by the ability of program owners to generate far more revenue per viewer from subscription sales. Home Box Office, for example, was very quickly able to outbid the big three television networks for the first television showing of theatrical movies because it earned so much more money per viewer. The added capacity and ability to identify and aggregate audiences may make it more profitable to provide many more types of programming on this basis.

Given the huge economies of scale, the most expensive programming will continue to be programming that attracts a mass audience. But subscription mechanisms may increase the profitability of such programming. Just as VCR rentals and cable premium channels displaced the television networks as the first means of showing theatrical movies on television, pay mechanisms may supersede advertiser-supported media as the first or exclusive option for traditional first-run network or syndication programming.

Narrowcasted Advertising. As a result of the development of more subscription and narrowcasted advertiser-supported programming, the advertising industry will undergo substantial retooling. The importance of advertising that is broadcast to national and local audiences will probably decline. More important than interactivity will be the probable tenfold increase in channels. This shift will require agencies to develop ads compelling to targeted audiences.[48] The availability of subscription programming may also be supported by the willingness of viewers to pay to avoid ads or see their favorite program when they want to or as soon as it is available. It might be possible, for example, for a viewer to watch a program without the ads by paying a nominal fee. Russell Neuman, a communications professor at Tufts University Fletcher School has said: "The people who buy BMWs, eat quiche, and drink white wine—the ones most advertisers want to reach—are those who can afford programs without ads."[49] One ad executive believes that mass marketing techniques will "go out the window" and that in an interactive world advertisers will switch to targeted marketing.[50] Advertisers may have to get into the programming business, including launching their own specialized networks, to be sure that their message is received by the right audience. Interactivity eventually might mean the launch of a McDonald's Network or an Exxon Network.

Personal Agents. As already noted, the cost of transacting business will fall, making it easier for consumers to interact with a vast audience very quickly. This interaction will dramatically improve the efficiency of the market for unique, rare, used, and perishable goods, where search and delay costs are high. When casting this wide net, a consumer will be able to create a "mesh" that will capture useful information but allow unwanted information to pass through.

Many industry observers believe that a form of software called *personal agents* could further decrease transaction costs. These agents could enable consumers to interconnect with other computers, making them even more efficient at sorting through shopping, entertainment, and other options. AT&T, IBM, and other leading companies are creating these cyber-valets that could find the proverbial needle of valuable detail in the haystack of today's information. One form of agent would have a more limited scope of operation.[51] Many companies are developing non-mobile agents. IBM, for example, is developing a software-based system called Intelligent Communications, which it envisions gathering a customer's voice mail, faxes, electronic mail, and online news reports. The system would sort this material by priority. To retrieve the data the user need only call it up with a personal computer, pocket phone, or personal digital assistant.[52]

AT&T, Lotus Development, and Novell are competing to develop much more ambitious mobile personal agents. AT&T's PersonaLink uses Telescript, a computer-programming language that allows an agent to move around a computer network until it completes the transactions assigned to it.[53] For example, one could forward a copy of all messages from Tom about Dick to Harry. Or one could send an agent to buy a TV, specifying the kinds of stores, model, price range, number of bids, and so on. The agent would travel to stores that have identified themselves as sellers of TVs in a PersonaLink directory. Having completed its mission, the agent would report back with the appropriate information.

Mobile personal agents raise several problems, however. First, they require more sophisticated technologies. AT&T and its partners have fallen behind in their plans to add agents to the PersonaLink service that can do more than relatively simple tasks, such as ranking incoming messages. Second, retail networks such as PersonaLink create the classic chicken-and-egg debate over who will sign up first, customers or merchants. Finally, there is the difficult problem of computer security. Some data suppliers may be wary of letting an agent roam through their computers in search of information the user has requested, out of fear the agent could spread a computer virus or steal information. According to Samuel May, an analyst at Yankee Group Inc. in Boston, "The owners of information repositories are scared to death about agents."[54]

Nonetheless, the promise for mobile agents appears great and the implications profound. If, as some industry observers believe possible, agent-based networks are accepted by 20 percent of the population, they could revolutionize consumer retailing. Shopping networks such as QVC envision consumers using customer profiles to provide cyber "heads-ups" on shopping bargains, new movies and books, and news on areas of special interest. These profiles could be generated by the customer or be based on the customer's preferences as revealed through usage.[55]

Smart Everything. Other machine-to-machine interactions made possible by the telecom revolution will allow consumers to operate, maintain, and repair major appliances and cars more quickly and at lower cost. Today it is difficult to buy a major household appliance that does not contain a microprocessor. In the not-too-distant future it will be possible to buy appliances with microprocessors that can be connected through a "wireless" network to a central server capable of diagnosing problems. A machine or expert could remotely analyze computer or car problems or anticipate them before they occur.[56]

Although the telecom revolution will have extensive ramifications for these and other industries, it is important to remember that its most profound ramification will be the discovery-process effect. The telecom industry is appropriately considered a primary driver of the Innovation Age, because by increasing the number of beneficial interactions it greatly enhances the market's discovery-process capability. This discovery-process effect will be powerfully felt throughout the entire global economy. Continuing improvements in digital technologies and the advent of full-blown competition in local telecom markets will only heighten the importance of the telecom revolution. Ironically, telecom is also one of the greatest opportunities facing public policy makers, because the benefits of the telecom revolution will be far greater if regulation in this area is reformed. As discussed previously, in the Innovation Age increased economic complexity and dynamism have changed the political economy of innovation. The insights of Hayek and Schumpeter can help us know what to expect and enable us to decide how policymakers should adapt current regulations to the coming situation. Chapter Five considers how telecom policy should be reformed to give Americans the full benefit of the telecom revolution.

Chapter Five
Telecom Reform

Discovery-Process Guide

The Telecommunications Act of 1996 signed into law on February 8, 1996 is the most important telecom legislation of the last sixty years. This legislation reflects the widespread recognition that a new federal regulatory framework more conducive to innovation is necessary, if not long overdue. While the new telecom legislation "fumbles" some key provisions and on others "punts" to the federal and state regulators, it sets a new legal and policy "playing field" based on the presumption that competition, open entry, and market forces should regulate the telecom industry.

The legislation preempts state and local barriers to the development of local telephone competition and requires local telephone companies to provide interconnection with competitors on reasonable terms. The Federal Communications Commission and state regulatory commissions must determine what the local telephone companies are to do to open their markets to competition. When this competitive "checklist" is met, the regional Bell companies will be free to enter the long-distance telephone and equipment manufacturing markets. At the same time, the FCC and states must reform universal service subsidies so that sustainable support mechanisms for universal telephone service will be funded in a nondiscriminatory fashion. Long-distance rates will have to be geographically averaged.

Under the Telecommunications Act, regulation of rates for many cable services is eliminated in small communities immediately and for all systems within three years after enactment. The FCC must abstain from regulating telephone companies if such regulation is not necessary to protect the public interest or consumers. Multiple ownership rules for broadcast radio and TV stations are relaxed. Television manufacturers are required to install "V-chips" in most TV sets to enable parents to screen out violent programming. Additionally, anyone "knowingly" transmitting information "indecent to minors" across a computer network can be jailed and fined.

In implementing the legislation, the FCC and the state public utility regulators will have to grapple with many difficult issues over the next few years. These issues will involve the three main categories of telecom regulation: (1) regulating monopoly, (2) assuring universal telephone service, and (3) managing the electromagnetic spectrum. Regulators will have to consider new approaches to these traditional regulatory concerns. For example, in the local exchange telephone market, where significant monopoly power still exists, the new legislation requires regulators to oversee the creation of conditions conducive to the development of local competition. Can policymakers devise new, flexible methods of accomplishing these goals? As to other monopoly-power issues, we need to determine when telephone, cable, and other companies should be permitted to merge in the same market. We also need to ascertain the dangers of private companies that achieve dominance through private and de facto standard setting in the unregulated marketplace.

Regarding the availability of telecom service, regulators will have to harmonize the goals of universal availability of telecom services with competition and efficient pricing. How can universal-service subsidies of residential rates be reduced without decreasing telephone subscribership? Under what conditions might universal service be expanded to include new technologies and services?

The allocation and assignment of the electromagnetic spectrum presents one of the most important opportunities for telecom reform. How can the current, federally-managed system be improved so as to generate the information needed to better meet the competing uses of spectrum across the country? To what extent can and should the management of spectrum be privatized?

These questions must be approached differently in light of the imperatives of the Innovation Age. As explained earlier, in an era of evergrowing change the opportunity cost of intrusive and dilatory regulation is perpetually increasing. As policymakers consider the many questions and set about reforming the three main areas of telecom regulation, the concepts developed by Hayek and Schumpeter offer a guide. The essential insight embodied in these economists' theories of spontaneous order and creative destruction is that the marketplace offers a superior process for discovering and utilizing the innovations that will most benefit society. In the Innovation Age, the prudent regulator will seek to build on the market's comparative advantage as a discovery process.

Exploiting this comparative advantage will require policymakers to rely more on market forces and use more de-centralized and less intrusive forms of regulation. As noted earlier, market prices and incentives efficiently reallocate resources in response to changes in consumer demand and technology; they also encourage risk-taking and innovation. Fostering competition, eliminating governmental barriers to entry, and protecting property rights and freedom of contract, help to keep the market's discovery process strong. Regulation necessary to check market failures can be adapted to help preserve the information-producing and feedback mechanisms of the marketplace. Explicit and competitively neutral subsidies will minimize distortive effects on the price system. These principles, which rely primarily on using the market's superior ability to discover what works, can help provide an approach for policymakers as they deal with the complex issues and questions ahead.

Monopoly Regulation

Market Power

A firm has market power when it can profitably restrict total output below levels that would exist in a competitive market. Conversely, in an effectively competitive market an individual firm cannot profitably affect output or price levels. The principal basis for telecom regulation should be to check the presence of monopoly or substantial "market" power. When a firm exercises market power, consumers lose out in two ways. First, consumers who continue to buy the affected product pay higher prices that are bid up from competitive levels. Second, consumers who forgo the product at the higher price are worse off because they lose the benefits of the product whose competitive costs they would be willing to pay.

In the Innovation Age, substantial market power will be harder to find, and as illustrated earlier in Table 1 of Chapter I, it will be potentially far more costly to correct. There are three major markets into which the telecom world can be divided: software, equipment, and transmission. We will now consider the current relative competitiveness of each of these markets.

Telecom software. Like personal computers, the fiber optic lines and set-top boxes of digital networks will require software. In addition to the operating system software, the digital network's "software" will consist of movies, TV programming, video games, data bases, video-on-demand, electronic shopping, and other information services. The software market, broadly defined, is highly competitive and has low barriers to entry.[1] Indeed, as we will see later in this chapter in the section on American opportunity, according to some observers the "winner-take-all" nature of these markets has increased the level of investment and competition in them to extraordinary levels. As for software that might be unique to the telecom revolution, powerful forces exist to check possible monopoly power in the Innovation Age. The economies of scale that exist in writing software, building specific databases, and marketing products nationwide can be expected to create incentives for firms to be the first to establish a strong market presence. Many will find it profitable to price their services low enough to become the dominant provider in a particular category and use promotional efforts to convince even the marginal, least interested users to buy their wares.[2] Moreover, the experience with computer software applications suggests that the information services market will continue to be highly competitive. Although firms that establish dominant positions early can be expected to earn lucrative returns, an ever-growing market, low entry barriers, and rapid technological advances will require firms to innovate continually.

Telecom equipment. Telecom services require that equipment be installed at the customer's premises. In the near future, in order to get many of the new digital services that will become available, consumers will have to buy set-top controllers for their TVs or PCs. Later, these set-top boxes may well be integrated into the TV or PC. Some observers contend that there will be large digital converters capable of managing all the digital appliances in a household. These so-called "digital furnaces" could manage electrical appliances while providing access to the new digital information services. These larger units may be able to save a consumer money by spreading the cost of the equipment over many intermittent uses. On the other hand, declining microprocessing costs and the advantages of flexibility and ease of use make the modular approach more plausible.

Whatever equipment configuration prevails, dominance will almost certainly be hotly contested in the marketplace in much the same way that the makers of PCs, workstations, and mainframes struggle for market position today.[3] There appears to be little reason for government regulation in this area. As discussed later in this section under *Setting Standards*, the many drawbacks of establishing mandatory standards in an innovative industry make it premature to speculate about the need to require an open standard for operating systems for these set-top boxes.

Telecom transmission. Today the transmission market is divided into two submarkets: the one-way video or television market, and the interactive (usually voice-only) market provided by telephone companies. These two markets are now beginning to merge. Long-distance telephone, cable TV, and cellular companies are launching an all-out attack on local telephone markets. The leading local telephone companies, for their part, will shortly be entering the long-distance telephone and local television markets. In some cases these new developments will serve to heighten competition that is already intense. In others, it will give consumers an effective choice of providers for the first time.

Open entry and competitive rivalry are already the order of the day in the television market. Transmission capacity is provided by terrestrial broadcasters, cable TV operators, direct broadcast satellites, wireless cable TV systems, VCRs, and telephone companies. Ninety-eight percent of all U.S. homes can receive the major TV networks. More than 90 percent of all households can subscribe to cable TV, and roughly 60 percent of all households do. In the space of a year, more than a million households have subscribed to DBS, obtaining access to an additional one hundred networks of programming and information. Eighty-five percent of all U.S. homes have VCRs. With all these choices available, local broadcasting markets are becoming more fragmented and losing audience share to non-broadcast programming. Cable systems, although they may have market power in providing certain cable networks, face competition from broadcasters, VCRs, and increasingly from DBS.

The telephone market is also currently divided into two segments: long-distance and local exchange. The long-distance telephone market exhibits intense rivalry. The three major carriers—AT&T, MCI, and Sprint—have frequently introduced new discount plans that cut prices. Customer turnover between the three major carriers has been quick and highly responsive to changes in price. In 1994, nineteen million customers (20 percent of all customers) changed carriers twenty-seven million times. At the same time, a significant group of smaller long-distance carriers grew to 17 percent of the market.[4]

Regarding local telephone markets, it now seems that breakthroughs in fiber optic and wireless technologies will make competition possible. Competition is already flourishing among firms called Competitive Access Providers, or CAPS, which use primarily fiber optic networks deployed in urban areas; these networks furnish the local-access component of long-distance calls to high-volume business customers. CAPS have been expanding rapidly in recent years in both large urban markets and smaller cities.[5] Currently, CAPs are providing an alternative only for lucrative, large business customers who generate a disproportionately large share of a telephone company's revenues. It is likely that in the near future large-scale joint ventures between long-distance companies, wireless providers, and cable companies will be targeting high-margin residential and small-business customers.[6]

For the vast majority of America's households with telephones—roughly 94 percent—there is no alternative to the local telephone company; but FCC staff reports conclude that the prospects for local competition are promising.[7] Cellular and new wireless services will soon provide a cheaper, high-quality alternative means of traversing the "last mile" to the customer.

Problem area. Notwithstanding these developments, a major obstacle to the growth of widespread competition in the local telephone market may still exist. It appears that the biggest question about the prospects of effective local telephone competition is whether wireless telephone providers will be able to interconnect with underlying infrastructure provided by someone other than the local telephone company. Today more than 90 percent of all wireless calls still interconnect with a wire-based infrastructure at some point. Without a competing wireline infrastructure, such as a local cable television system, the local telephone company will retain bottleneck leverage over almost all local calls—preserving most of its monopoly power.[8]

Given the huge economies of scale in fiber optic networks, the prospects for viable alternatives to the local telephone company's core infrastructure might seem questionable. Questions, in fact, have been raised not only about the ability of telephone companies to profitably provide video on existing copper plant but, more fundamentally, on the wisdom of telephone companies building another broadband network.[9] However, the coaxial cable of cable TV systems is comparable to fiber over short distances, and high capacity cable modems and wireless technologies such as PCS promise attractive ways to access customers.[10] Although cable operators may face high costs, technical integration problems and customer resistance in providing telephone service,[11] the infrastructure is largely in place, which makes it likely that local telephone competition will get a try-out.[12] At a time when voice and video markets appear to be merging, it might eventually prove more difficult for telephone companies to enter the local cable television market than for cable TV companies to enter the local telephone market.

Regulating Monopoly

Even in the two telecom markets where there is evidence of market power—some cable markets and local exchange telephone markets—federal regulation needs to be exercised prudently. Experience with federal regulation of cable rates indicates that it has been largely counter-productive. Recent research by economist Thomas Hazlett shows that even though cable systems had significant market power, cable prices, adjusted for quality, went down after federal *deregulation* became effective in 1986. Although nominal cable prices actually increased, more households subscribed to cable because deregulation led to valuable investments in new cable programming.[13]

Conversely, *reregulation* of cable rates pursuant to the 1992 Cable Act appeared to curtail cable companies' investment in new programming services. Many cable operators declined to take on new channels in the year following the FCC's reregulation of cable rates. Brian Lamb, CEO of the Washington-based public affairs network C-SPAN, said of plans to distribute a new channel, "we're dead in the water" without an FCC rule change.[14] The FCC has since modified its price cap regulations to give cable operators greater incentives to carry new cable channels.[15] Despite the change, executives for one cable company announced that it was ceasing operation as a 24-hour network, labeling the FCC action "too little, too late."[16]

In the local exchange telephone markets, where the existing telephone companies appear to have substantial market power, regulation is generally better left to the states, subject to federal guidelines to assure that conditions conducive to competition are preserved. Devolving regulation of local telephone facilities to the states would eliminate needless complexity arising out of federal/state "separations." Today the FCC and the states are involved in an elaborate process of "separating," i.e., assigning a portion of local telephone service costs to the interstate jurisdiction to cover the local "access" component of long-distance calls that cross state lines. The FCC then oversees the access prices that interstate users pay to cover these costs. There is little or no economic justification for subjecting any particular level of these costs to regulation by a federal agency. This unnecessary complexity could be eliminated by giving state regulators authority to regulate most or all of these costs. Regardless of what happens to the assignment of costs between state and federal jurisdictions, states should have more freedom to try different rate structures that they believe might be more efficient and accountable to local desires. Unlike state limitations on entry—which are preempted by the new federal telecom legislation—there is a potential benefit from state experimentation with different rate structures and regulations designed to promote competition.

Although such devolution would give state regulators new responsibilities, especially with regard to implementation, important federal guidelines can be designed to protect essential national interests, such as promoting interstate commerce and fostering competition and innovation. While experiments in states such as California, Illinois, Massachusetts, Michigan, and New York should inform the FCC's regulatory efforts, the FCC should use its authority under the 1996 Act to achieve three objectives. First, alternative local exchange competitors should get access to unbundled elements of local telephone networks at forward-looking, cost-based prices. Effective resale will foster entry by competitors and promote efficient repricing. Second, whatever local exchange costs are allocated to the interstate jurisdiction, they should be recovered efficiently. Reform of these access rates will foster the efficient use of long-distance networks. Third, states should be required to recover any universal service subsidies in competitively neutral ways, without discriminating against alternative local exchange competitors or interstate users of local exchange networks. These points are discussed at more length later in this chapter in the section on "Universal Service."

Beyond the requirements of federal guidelines, states should aggressively move to adopt efficient repricing and market-oriented regulation. Most states have already abandoned traditional rate of return ("cost-plus") regulation, but substantial reforms are still called for. In addition to the rate rebalancing discussed below, local telephone companies will need pricing flexibility to respond to the competitive offerings made by new entrants. In other industries, failure to allow such price flexibility led to unfortunate results.[17] Although the full relevance of the computer industry's experience to digital telecommunications remains to be seen, policymakers would be wise to heed its lessons and avoid traditional forms of regulation that could stifle the innovative potential of digital technologies.

Antitrust Enforcement

In the Innovation Age, changing technology, relaxation of regulatory barriers to entry, and uncertain market impacts make a case-by-case approach to antitrust enforcement prudent. Guarding against a substantial lessening of competition in relevant antitrust markets through industry-wide structural rules that prohibit mergers no longer works when market boundaries are changing and generally expanding. Often markets themselves will combine, reducing market concentration as existing companies are forced to operate in a bigger pond.[18] In other instances, mergers will permit firms to compete more effectively. The new federal telecom legislation makes a good first step in this direction by liberalizing existing ownership limits: a long-distance company can now merge with a regional Bell company; local telephone and cable companies can merge in small markets; and in broadcast markets, one company can acquire TV stations reaching 35 percent of the nation's households and an unlimited number of radio stations nationwide. Experience with these relaxed standards, together with merger review by the Antitrust Division of the Department of Justice and the Federal Trade Commission, may well encourage granting further relief in the future.

A striking example of the need for regulators to be cautious is provided by the failed cable/telephone mergers of 1993. Over the course of 1993, three major mergers between local phone companies and cable TV companies were announced. The first proposed alliance was US West's investment in Time Warner. US West, a regional Bell company[19] and the dominant local telephone company in the north central and western United States, purchased 25 percent of Time Warner, a diversified communications firm with holdings in cable TV, magazines, and movie studios. The second and largest of the proposed mergers was between Bell Atlantic, the dominant local phone company in the mid-Atlantic states, and Telecommunications Inc. (TCI), the nation's largest cable company. The last proposal was a merger between Cox Enterprises and Southwestern Bell Corporation (SBC). Cox Enterprises is also a diversified communications company with holdings in cable TV and publishing, while SBC is the dominant local telephone company in the southwestern United States.

Announcement of the mergers fueled speculation that competition in local telecommunications services was just around the corner, and that new services—such as video-on-demand—would soon become reality. These mergers were viewed as the means to unite the video capabilities of cable TV and the interactive capabilities of a phone company's switching technology. In addition, such mergers were expected to facilitate competition in local telecommunications services. The merger prospects generated enormous excitement on Wall Street. By mid-1994, however, the movement toward a unified telecommunications industry had stalled, and the two largest proposed mergers had been canceled. Industry analysts were left wondering why.

Some analysts attributed the cancellations to investor disillusionment with the mergers themselves. Others argued that governmental developments, such as the reregulation of cable prices, might have affected the prospects for the proposed mergers. While details of the mergers were being worked out, rumors had begun to circulate that the FCC would be strict in its reregulation of the cable TV industry as mandated by the Cable Act of 1992. These rumors were eventually confirmed by the FCC's February 1994 regulatory announcements that rolled back cable prices substantially. The cancellation of the Bell Atlantic/TCI merger and the dissolution of the Cox/SBC venture followed. Ray Smith, the CEO of Bell Atlantic, has stated that the resulting reduction in TCI's cash flow dramatically reduced the value of TCI stock and effectively killed the deal.[20]

Detailed analysis of the behavior of telecommunications stocks surrounding the merger announcement and cancellation dates indicates that the telephone and cable TV mergers of that year were likely to have spurred competition in local telephone service.[21] The rise in value of the long-distance and cellular telephone companies following the merger announcements suggests that these mergers would have enhanced competition, not market power, in local telecom markets.[22] After the TCI/Bell Atlantic merger was canceled, John Malone, TCI's CEO, stated that the cable/telephone mergers were motivated by the desire of the relevant regional Bell companies to compete in other local telephone companies' markets.[23] Also, the valuations investors placed on the merging companies support the view that any gain to the merged companies from increased efficiencies might well have been offset by the negative effect on those companies of enhanced local telecom competition. If regulation played any role in the cancellation of the mergers, it is extremely ironic, because, as mentioned, analysis of stock prices suggests that these mergers would have benefited local telecom competition perhaps to the detriment of the merging companies themselves.

In the proposed cable/telephone mergers described earlier, the suspect regulation was cable rate regulation, but the lesson applies with equal force to antitrust enforcement in general.[24] Mergers that might have been problematic in the past may now be permissible because of their likely procompetitive effects. For example, so-called cluster mergers of cable systems in local markets can reduce the cost to the cable companies of building a wireless telephone network that covers an entire local telecom market. That reduction would, in turn, improve the systems' ability to compete against existing telephone companies. It is also possible that allowing cable/telephone mergers may foster the deployment of advanced technology in markets where dual wires to the home or even the neighborhood may not be feasible. Likewise, the recently proposed mergers between regional Bell companies—Bell Atlantic with NYNEX and SBC with Pacific Telesis—could provide needed financial clout and thereby promote competition in the global telecom market or in domestic voice and video markets.[25]

Setting Standards

In many quarters, support is growing for government regulation of technical standards in the telecom market, as a means of fostering competition. Close analysis of the computer market, however, demonstrates that this would be a mistake. As noted earlier, the computer industry has achieved the advantages of standard architectures, interoperability, and interconnectivity in the marketplace with minimal governmental intrusion and has preserved market incentives that promote innovation. Because existing architectures can be challenged, hardware manufacturers are under constant pressure to revise or even cannibalize popular architectures to remain competitive. The same market pressures from consumers and software companies for uniformity and "coopetition" will operate on the companies that provide telecommunications "hardware" and "software." At the same time, the advantages of private standard setting and technical flexibility are also likely to be great, because digital networks will experience rapid technological change for the foreseeable future. Firms that fail to adopt new technologies and exploit them to their advantage will be threatened by firms that are willing to make the necessary sacrifices and investments to exploit new technologies. This will hold true regardless of whether the technology that underlies the networks is open or proprietary.

The imposition of mandatory government standards, however, could stifle innovation, because technical change would require regulatory approval. This process invites abuse by concentrated economic interests that wish to resist change to preserve their market shares. Government standard setting often leads to the adoption of a least-common-denominator approach and places insufficient weight on important economic tradeoffs raised by technical decisions. Often, permitting variation and experimentation in the marketplace provides a more reliable basis for technical convergence. It is evident that the marketplace has successfully set standards for VCRs, CD players, and audio cassettes. Moreover, in some cases, complete uniformity may not always be desirable. For example, the availability of competing word-processing packages with varying degrees of complexity better meets differing consumer needs, and helps to spur innovation and price competition.

Given the high degree of uncertainty and the great potential for rapid innovation, such regulatory mistakes could be costly. Standards must be dynamic and responsive to changes in technology. When technology advances too quickly for government regulators to keep up, as is often the case, stagnation and inefficiency result. Consider the example of the conventional (NTSC) television standard. It has been changed only once in the last fifty years—to include color.[26] Meanwhile, any other would-be quality improvements inconsistent with the standard were unavailable to consumers.

Private standard-setting activities in digital telecommunications are already achieving the benefits of uniformity, relying on sounder information, and preserving incentives for innovation. For example, concord has been reached between contending potential producers of digital videodisks (DVDs).[27] Similar in size to CD-ROMs, DVDs hold seven to twenty-five times as much information and in a few years could replace videotapes, audiotapes, and CD-ROMs. Philips Electronics and Sony on one side, and Time Warner and Toshiba on the other, have agreed to a common format which should make sales possible as early as fall of 1996.

Even where the government has considered setting a standard, as in the case of HDTV, primary reliance on private efforts has been beneficial. With only limited help from the FCC, the leading proponents of digital HDTV systems have agreed on a standard. In addition to achieving uniformity, this private standard reflects the affected companies' assessment of the relative strength of their respective technical achievements.

Elsewhere, private efforts are under way to set important digital standards. Several companies are attempting to develop "set-top boxes" which receive and manage digital signals that arrive over a digital network. Companies such as Apple and Microsoft are trying to develop operating systems to run these set-top boxes and the services that flow through them. Following their own success in computer operating systems, both companies would like to capture this market.

Under the new telecom legislation, the FCC must ensure the commercial availability of converter boxes and other multichannel consumer equipment, although the Commission is instructed to avoid chilling the development of new equipment.[28] It can, however, rely on industry standards groups to determine what, if any, standards are needed. In the past, some companies interested in developing applications software and making the hardware have wanted the FCC to require open specifications for these boxes.[29] Such governmental action is premature. Governmental mandates could undermine investment incentives and probably substantially delay the development and deployment of the technology. Although the FCC might serve a useful facilitating role, its imposition of a standard might cause costly mistakes or lock in one particular technology. The result could be a long-term loss of innovative services.

Universal Service

Two Rationales

Harmonizing universal telephone service with competition and innovation presents one of the most difficult challenges facing telecom policymakers. As discussed earlier, new competition in local exchange markets will first attack the source of today's system of telephone subsidies. Most of the subsidy amounts are raised by inflating the price long-distance callers implicitly pay for accessing local telephone networks used in originating and terminating their calls. Many local business services are also priced above cost.

Without reform, inefficient deployment of facilities will be likely in high-margin services, because new entrants will base their investment decisions on inaccurate price signals. Today's subsidies also discourage the efficient deployment of cost-saving technologies in rural areas. These problems are further complicated by political pressures, evidenced in the new telecom legislation, to expand existing subsidies as a way of assuring that new technologies and services are universally available. Such premature calls for redefining universal service can increase the risk of deploying new technology and services everywhere. To understand how universal service goals can be maintained and advanced, we will first consider their origin and rationale and then propose reforms based on experience with initiatives at the federal and state levels.

Neither history nor experience provides a clear definition of "universal service." The phrase appears nowhere in the Federal Communications Act of 1934, and its original meaning is in doubt. The term was apparently coined in the early 1900s by Theodore Vail, then CEO of AT&T, who believed that "some sort of a connection with the telephone system should be within reach of all." The current definition of universal service—that affordable telephone service should be available to all—may date back only to the 1960s or 1970s.[30] In the Telecommunications Act of 1996 there are six "universal service principles," one of which is that rural customers should have access to services available in urban areas at reasonably "comparable rates;" a final catchall principle allows the FCC, together with selected state regulators, to add other principles consistent with the public interest and the 1996 Act.[31]

Although many people speak as if universal service has been achieved in the U.S. for decades, this determination is subjective. According to a recent FCC report, as of July 1995 more than 94 percent of U.S. households had a telephone and nearly 95.3 percent had access to a telephone. In 1983, the FCC found those measures to be 91.4 percent and 93.8 percent, respectively. In 1961 the penetration figure was roughly 80 percent.[32] Telephone penetration rates, moreover, vary significantly along income lines. For example, in March 1993 households with less than $5,000 in annual income had a telephone penetration rate of only 72.9 percent. For households with incomes between $15,000 and $19,999, the figure was 93.6 percent, and for households with incomes over $60,000, telephone penetration topped out at 99.6 percent.[33]

Analysis of the two public policy rationales for "universal service" provides some clarification of what this term has meant thus far. First, most Americans regard telephone service as a necessity, if only for health and safety reasons. From this fact, many federal and state regulators have long assumed that telecom regulation should, separately from general welfare assistance programs, assure that every American household has access to affordable telephone service. Of course, the prices of many other consumption items commonly viewed as necessities—food, housing, gasoline, and transportation, to name a few—are not regulated to favor residential customers over business customers or to even out price differences that might exist between urban and rural locations. Although the foundation of the necessity rationale for subsidized access is debatable, universal service policies tailored to assist only low-income *households* would dramatically reduce subsidy amounts and pricing distortions that exist under the current system, and would actually increase the percentage of American households with a telephone.[34]

The second rationale for universal service rests on the rather special policy considerations raised by the nature of telephone networks. Subsidies that encourage non-subscribers to buy telephone service can also be justified if they are offset by the increased value of phone service to all other callers who now have more users to call or be called by. This rationale rests on the possibility that some of the benefits to society from having an individual subscribe to the telephone network are external to that individual's calculation of costs and benefits. Theoretically, it is possible that subsidizing telephone access would be efficient because an individual's decision on whether to subscribe may not take into account the value to others of his own presence on the network.[35]

This "externality" argument could apply to all areas of the country, but it is thought to be particularly important to residential callers in rural areas, given the higher costs of service and lower average incomes found in those areas.[36] As long as penetration rates remain high, however, the question of whether basic access rates should be subsidized on externality grounds remains an academic question. Consequently, the continuing need for universal service subsidies has been gauged by estimating whether and to what extent consumers would drop off the network if the price of telephone access increased significantly. The overwhelming preponderance of the evidence indicates that the demand for access to the local telephone network is highly insensitive to price changes and is becoming more so over time.[37]

Thus, the concept of universal service has had two policy underpinnings: (1) telephone service is a necessity that should be available to all, and (2) subsidy of telephone rates is justifiable on efficiency grounds since everyone benefits from expanding the reach of the telephone network. These two rationales for universal service are based on universal subscribership, not maintenance of low prices in some relative or absolute sense. With the achievement of high household penetration rates, however, regulators have transformed universal service into a policy of simply keeping residential telephone rates low, especially in rural areas. By establishing as a universal service principle the idea that rural rates should remain reasonably comparable to those in urban areas, the new legislation apparently has added to the problem. In fact, the legislation gives regulators substantial flexibility to reform the current subsidy system. For example, regulators could add universal service principles such as "the promotion of competition," "the deployment of advanced technology," and "local accountability." In this way they could justify implementing policies that would efficiently keep household penetration rates high, but with minimal intrusion on the development of competition and new digital services.

Experience shows that reducing subsidies need not affect telephone penetration rates. In the 1980s reductions in the long-distance subsidy of local residential rates through the imposition of the so-called "federal subscriber line charge" were a huge success. Under this program each residential subscriber currently pays $3.50 per month to cover a portion of his basic subscriber line costs that do not vary with usage. In the past those local telephone network costs were covered not by the local callers themselves but by interstate long-distance callers, who paid higher prices. Although from 1984 to 1990 telephone rates went up moderately,[38] telephone penetration increased from 91.4 percent to 93.3 percent; ten million additional households subscribed to telephone service; and the number of households without telephone service decreased by 1.1 million.[39] At the same time, the subscriber line policy cut long-distance prices approximately 40 percent, which stimulated enormous increases in long-distance calling and annual consumer welfare gains amounting to billions of dollars.[40]

These lower long-distance rates also benefited many rural and low-income customers. Because rural residential callers make more long-distance calls on average than do urban residential customers, they reaped substantial benefits from the long-distance price decreases.[41] One study found that low income (lifeline) customers paid more than 60 percent of their monthly bill for long-distance services[42] and that between 1984 and 1990 the decrease in interstate long-distance prices had a positive effect on penetration of approximately three times the magnitude of the increase in basic exchange access prices.[43]

Proposed Reforms

Reform of universal service policies will require both procedural and substantive change. In large measure these issues should be devolved to the states, but the new telecom legislation appears to require a substantial role for federal subsidies. Regardless of which jurisdiction is involved, however, the subsidies need to be targeted and reduced. By better aligning prices with costs, further reductions in the total level of subsidy would produce substantial long-distance price decreases, stimulate long-distance calling, and spur the efficient development of competition in local telephone markets.

Federal Policy. The FCC should use the latitude that the Telecommunications Act of 1996 affords it to make important reforms necessary to promote competition and innovative technology.[44] In general, any federal universal service system should use market mechanisms to limit its costs and strive to target subsidies to those households that need them to stay on the network. Blanket subsidization of residential rates in high cost areas is inefficient and unfair. For example, one telephone company is proposing to use universal subsidies to bring telephones to a few dozen expensive summer homes at a mountain-lake resort near Zion National Park at a projected cost of $430,000.[45] To minimize such abuses, subsidies should be targeted to poor households. The goal of aid to high-cost areas should be to keep rates affordable as measured by household penetration rates. Thus, the term "comparable rates" in the Act should be interpreted liberally.

While the new legislation requires that so-called "eligible carriers" be compensated for the high cost customers they serve, it does not preclude a system that targets and reduces subsidies as long as rates remain affordable. For the sake of illustration, in most of the country the monthly cost of providing telephone service is roughly $18 per month or less. In areas with fewer than one hundred people per square kilometer, that monthly cost rises to $37. In areas with ten or fewer people per square kilometer the cost rises to $65 per month.[46] A more equitable approach would be to give each eligible carrier a subsidy equal to the number of high cost lines it serves times a subsidy per line that reflects each line's supra normal service costs (e.g., $37-$18=$19). The subsidy-per-line amount should then be reduced each year over a ten year transition period. Monitoring household penetration rates could serve as a failsafe during the transition to more targeted subsidies.[47]

Finally, funding mechanisms should not favor one technology or industry, or unnecessarily suppress demand. They should be explicit and accountable to the public. Ideally, they would come from an explicit and transparent charge paid by all telephone customers or all telecom providers. An increase in the federal subscriber line charge levied equally on all business and residential lines would be an efficient means of funding any federal universal service subsidy.

State Policy. Consistent with the new telecom legislation, the states should be given significant responsibility for determining universal service policies. Consistent with the competitive neutrality principle and other federal requirements, the separate states could choose different methods of paying for universal service. For example, by mandating local measured service (LMS) options and other means, a state might be able to recover the subsidy costs in efficient and equitable ways. Under LMS, callers get cheaper access and initially pay higher rates for actual calls made. At higher volumes of calling, the usage rates can taper off sharply. Although the implementation costs can be significant, LMS is an efficient and sustainable means of recovering local networks costs because the marginal (usage) price of basic telephone access for large users is low. Consequently, they are less likely to leave the network or suppress usage.

Importantly, this approach appears to moderate the impact of higher rates on those individuals who might otherwise drop off the network. For example, one study found that where LMS was available (over half of all households reside in areas where LMS is available), changes in flat rate prices had no effect on penetration rates.[48] Another study found that if local rates were to double, adding a measured rate option would add three percentage points to overall household penetration rates among low income households; adding a service option targeted to low income households (so-called lifeline service) would add another five or six percentage points.[49]

Regulatory reforms undertaken in Illinois also demonstrate that efficient pricing of local telephone service need not undermine the goal of universal access.[50] From 1986 to 1993 the Illinois Commerce Commission (ICC) restructured residential telephone rates, phasing in mandatory local measured service throughout the state. During the same period, the access line price under LMS to subscribers in the various Illinois calling regions was increased roughly 10 percent to 30 percent. With the implementation of state-wide price restructuring in Illinois, the actual penetration levels reported for the state by the FCC fell by a statistically insignificant amount—from an annual average 94.3 percent in 1990 to 93.6 percent in 1993. This measured decline in household penetration for Illinois is three to nine times less than would be expected given the accepted range of demand elasticities for telephone access.

Several factors may explain this disparity: telephone access prices that are low in absolute terms under LMS, reduced long-distance rates,[51] and an increase in the demand for complementary services available through the telephone network. Whatever the specific reasons, however, the Illinois experience demonstrates that telephone rates for residential customers can be restructured to reduce subsidies with minimal impact on universality of service.

Subsidy Expansion

Along with the reforms just suggested, the states should have the principal role in defining universal service to permit experimentation and allow for accountability to local desires. Thus, one state might wish to define universal service as the availability of a dial tone. Another state might want to expand it to include new interactive services. Yet many members of Congress, the Clinton administration, and various public interest groups have attempted to expand, rather than refocus, federal universal service policies. Some have even asserted that it should be made possible for all households to have access to digital service within five years.[52]

The new telecom legislation delays consideration of this issue, but it allows a future FCC to expand universal service policies for telephone usage to include new digital services such as interactive video. The telecom legislation directs the FCC to consider the extent to which new services are essential to education, public health, or public safety and whether "a substantial majority of residential customers" subscribe to them.[53] The legislation also requires companies to provide subsidized computer linkups to schools, nonprofit medical facilities, libraries, and some other public buildings.[54]

There is little basis for expanding the definition of the universal service subsidy, especially on the federal level, until more is known about the demand for and cost of the new digital services. Neither of the universal service rationales used to explain subsidies to local telephone networks justifies the proposed expansion of universal service to include new digital services. By no stretch of the imagination can the added benefits of interactive video services be regarded as necessities. Nor does the externality rationale for telephone universality apply here. For the foreseeable future, the benefits of the digital network to individual customers will probably derive from their ability to interact with large, centralized video and information data bases, not with other subscribers.

Some consumer advocacy groups contend that the current interactive business trials are a form of "electronic redlining." In fact, the evidence is that many of the business trials target representative samples of the affected communities.[55] Moreover, even if these new services are initially more successful among affluent consumers, this fact alone should not surprise or distress policymakers. For example, computer penetration rates are higher in high income households (whereas in 1993, the national household penetration rate was roughly 33 percent, for households with incomes above $100,000 the figure was 70 percent).[56] Survey evidence on the use of the Internet also indicates that the majority of users are upscale, affluent young professional men.[57] But as Richard Solomon, adjunct professor at NYU, notes, "all new technologies develop unevenly."[58] He adds, "Despite universal access today, it took about sixty years before two-thirds of the population had telephones."[59] It is worth noting that delay of that kind would not be expected in the Innovation Age.

Even for schools and libraries it is important that subsidy programs be undertaken with great care. There are several reasons for proceeding slowly and incrementally in this instance. First, these decisions will best be made on the local level, where there is more knowledge of where additional funds can be spent most effectively and what the value of using these new technologies might be. Although there appears to be significant success in at least one instance,[60] there is also evidence that the process of wiring schools will be expensive and slow. For instance, Texas spent $6 million between 1989 and 1993 to give just 5 percent of its classrooms access to phone lines.[61]

More fundamentally, as noted earlier, no one knows what the prospects for interconnected computing truly are. Building and providing services for use on this new infrastructure are inherently risky, and prematurely requiring an expansive vision of universal service could increase these risks—delaying or killing innovative efforts. We need to know more about the demand for and cost of these new services before expanding universal service. What we do know is that competition in local telecom markets will require carefully crafted changes in regulation to reform existing universal service policies. This may not be as glamorous as pushing visions of an information superhighway, but it is where policymakers can currently do the American public the most good.

Spectrum Management

Systemic Failures

The third main set of telecom policies in need of substantial reform comprises those governing the use of the electromagnetic spectrum. The electromagnetic radiation of particular frequencies or wave lengths provides the basis for radio and television broadcasting;

mobile cellular telephone, telephone, and data links; and satellite video and voice communications. The telecom revolution has heightened the importance of implementing long-overdue reforms that would make it easier to use spectrum in innovative applications.

Until now, spectrum management decisions have largely been made by the FCC. First, the FCC has allocated large blocks of spectrum to particular uses, such as television and cellular mobile telephones. Second, the FCC has assigned licenses to particular individuals, using administrative hearings or lotteries.

Ascertaining the most efficient uses of spectrum is an immensely complex task that requires vast amounts of continually updated information. Markets, as discussed earlier, "parallel-process" information gathered by millions of businesses and consumers; parallel processing then provides continuous price signals on the relative usefulness of the myriad competing demands and uses of spectrum. In the case of spectrum, the information required to make the necessary tradeoffs is greater than in most other areas of the economy, because the pace of technological development has created a raft of new uses for "wireless" communications and new ways in which spectrum can be efficiently deployed. The discovery-process capabilities of the market are uniquely suited to deal with this mass of information. The administrative process, on the other hand, cannot help but be overwhelmed by it.

Experience shows that the FCC has consistently made decisions that misallocated spectrum, created inefficient industry structures, and locked in outdated technical standards. For example, in the 1950s the FCC dramatically underallocated spectrum to VHF television, thus reducing the availability of television channels and impeding the development of a fourth national television network. In the 1970s it long delayed and then underallocated spectrum to cellular mobile telephony, creating only two cellular companies. (Such decisions have been incredibly costly—the cellular delay alone may have cost Americans 86 billion dollars.)[62] Time and again the FCC imposed a nationwide grid that did not account for local differences—for example, underallocating forestry spectrum in Idaho, overallocating it in New York City, and vice versa for taxicab spectrum. The FCC also locked in an analog technical standard for cellular telephony that was twenty years old and out of date by the time the FCC got around to assigning cellular licenses.[63]

Evidence shows that a second failure of the current system of spectrum allocation is its susceptibility to capture by special interests that are opposed to innovation and competition. In the face of political pressure, the FCC has repeatedly delayed providing spectrum to innovative new services. Past examples include Direct Broadcast Television, "wireless cable," low-power TV, and new FM radio and VHF TV channels. A long-running current example is the allocation and licensing of a satellite-delivered radio service to cars, a process that has dragged on for more than six years.

Seemingly, few companies and trade associations can resist the temptation to use the regulatory process to impede competition or innovation. In practice, the "public interest" standard, under which the FCC manages spectrum, has been used by private interests as a weapon against consumers. In this arena, H.L. Mencken's admonition applies with full force: "It may be a sin to think evil of others, but it is seldom a mistake." Regulated companies frequently have embraced special public-interest obligations in exchange for protection from competitors and innovation. Over-the-air television broadcasting is the best example. In exchange for promising to meet unspecified public interest obligations, TV broadcasters were given significant protection from competition from cable and DBS systems. Using Judge Richard Posner's term, this "taxation by regulation" on the part of unelected regulators, followed by the inevitable *quid pro quo*, has been extremely costly to consumers. The sad reality is that in this "regulatory game" the established concentrated economic interests typically triumph over the amorphous and diffuse interests of consumers.

A third failure of the current system of spectrum management is the overallocation of spectrum to the government sector. Although there is some debate over how much spectrum is effectively denied to the private sector, there can be little doubt that it is, in absolute terms, a huge amount. Table 2 indicates the amount of spectrum available for government access.

Table 2
Spectrum Allocation Below 30GHz (NTIA Data)

National Table of Allocations	*MHz*		*%*	
Government Exclusive	2,271		7.6	
Non-Government Exclusive	8,961		29.9	
Shared	18,768		62.5	
Available to private sector		11,167		37.2
Not available to private sector*		7,601		25.3

*NTIA's estimate of the shared spectrum without specific FCC service rules.

Of that spectrum in the usable range (roughly below 30 GHz), almost a full one-third (or 10 GHz) is unavailable to the private spectrum. Today government users of spectrum have access to spectrum at far less cost than nongovernment users do. For instance, in the recent auctions of spectrum for PCS, a new mobile service similar to cellular mobile radio service, the winners paid more than $18 billion for 90 MHz of spectrum. As Table 2 shows, the government has exclusive or "shared" access to thousands of MHz of spectrum. While many governmental uses of spectrum are vital, others are not. All of these uses should be justified economically. Most non-government experts in this area believe that government spectrum users do not use spectrum efficiently if only because they do not face market discipline or enjoy access to capital markets that would enable them to deploy new radio technologies. Inefficient use of this spectrum obviously entails a huge economic loss to the American public.

Spectrum Markets

The solution to the current spectrum gridlock is to give existing and new spectrum users more operational and technical flexibility to use their spectrum in the way they see fit so long as they do not interfere with their "neighbors." This "privatizing" of spectrum allocation decisions would obviate the need for an allocation system run by the FCC. The logic of this reform is easily illustrated. Under the current system a licensee has two basic choices; it can (1) use its spectrum for a narrow purpose with the specific technology that the

FCC has designated, or (2) give it back to the FCC. Given these choices, the current spectrum use will be economical to the licensee as long as the spectrum has any positive value at all. In economic terms the "opportunity cost" to the licensee, i.e., the value of the spectrum in alternative uses, is zero. The cost to society, of course, is the most valuable use of the spectrum that is foregone.

Once a licensee is given the freedom to use its spectrum across a broader range, however, the licensee has a strong economic incentive to consider the relative merits of alternative uses. The more flexibility the licensee has, the more likely that its use will be the highest and best use, reflecting the spectrum's true opportunity cost to society. This principle of spectrum reform can be applied to both "new" and "old" spectrum.

New Spectrum. First, the FCC should make unassigned private sector spectrum and government spectrum available to private users. The spectrum should be structured in highly flexible parcels with minimal government constraints. These parcels should be given the legal protection and freedoms that attach to real property—i.e., technical and operational freedom, protection from trespass or unlawful takings, and the ability to assign, lease, or sell. (Of course, these new users should be subject to antitrust and other laws that apply to businesses generally.)

In addition, where there are mutually exclusive bids for these new parcels—in other words, where there are more applicants than available licenses—winners should be selected by auctions. Auctions are the most efficient means of assigning licenses for three reasons. First, they will make assignments more quickly. The two-year delay typically imposed by administrative hearings comparing applicants[64] (or comparable delays for lotteries where thousands of parties apply to win a license) can easily wipe out a large part of the net present value of a license. Second, by assigning licenses to those who are willing to pay the most for them, auctions will on average ensure that licenses go to those parties most likely to put them to their highest valued use. Third, auctions will make these assignments at less cost to society. Instead of using scarce resources for expensive lotteries and comparative hearings, applicants compete by adding zeros to the size of the transfer payment they are willing to make to the government.

Table 3 shows that as of May 1996 the FCC had raised over 20 billion dollars in auction revenues.[65]

TABLE 3

Auction	Number of Bidders	Number of Licenses	Total Amount of Winning Bids* (millions)
Nationwide Narrowband PCS	29	10	$617
Interactive Video and Data Services	472	594	$214
Regional Narrowband PCS	28	30	$395
Broadband PCS A and B Block	30	99**	$7,034
Direct Broadcast Satellite	3	1	$683
Direct Broadcast Satellite	2	1	$52
Broadband PCS C Block		493	$10,200
Multipoint Distribution Service		493	$216
900 MHz Specialized Mobile Radio		1,020	$204

*Based on net revenue figures compiled at the conclusion of each auction from FCC press releases.
**Three additional licenses, valued at $702 million, were awarded to pioneer's preference winners.

Where there are mutually exclusive applicants, auctions will be a highly efficient way to raise government funds. The primary value of auctions, however, is not that they raise money for the government. Care should be taken that auctions do not become an obstacle to the principal goal of reducing spectrum scarcity. Spectrum should be released as rapidly and efficiently as possible, even if doing so reduces the government's auction proceeds. This is because the overriding goal of spectrum reform is to free up spectrum for the benefit of the American economy.

Old Spectrum. The second way to achieve spectrum reform is to deregulate or privatize existing spectrum uses. As with the newly assigned spectrum, Congress should move to give existing users full flexibility to use their spectrum, subject to interference and international treaty constraints. Licensees should be encouraged to negotiate with their "neighbors" to reach mutually beneficial agreements on possible new uses of their spectrum. This approach creates a private incentive to "reallocate" spectrum and challenge the vested interests that can be counted on to oppose new competition and innovation. Likewise, government agencies should be encouraged to sell or sublease their spectrum to the private sector.

Those who see privatizing existing spectrum as a government "giveaway" miss the point. The existing users have for the most part purchased their spectrum in the resale market. It is not efficient, fair, or politically feasible to require them to pay for their spectrum again. Moreover, top-down reallocations of spectrum by the FCC will be difficult because the existing users will collectively oppose the release of hundreds of MHz in flexible allocations. Privatizing existing spectrum, by contrast, will create a competition among existing licensees to move "beachfront" spectrum out of its current "landfill" uses. In this parlance, the current "beachfront property owners" may not appreciate these efforts, but they will not be able to stop them. Indeed, they might well choose to join them. In the end, America's "beach-consuming public" will benefit.

The Telecommunications Act of 1996 has made modest improvements by lengthening broadcast license terms to eight years and strengthening licensees' legitimate expectations for license renewal. Also, during the past fifteen years, the FCC has increasingly recognized the concepts of flexibility and privatization in its spectrum-management decisions—for example, giving cellular and direct broadcast licensees substantial flexibility and defining the new personal communications service very broadly.

The time has come, however, to speed up this process dramatically. A centrally planned, government-run allocation system probably never made sense; it has cost Americans billions of dollars in foregone wealth and prosperity. Now, the accelerating pace of technological change and the increasing complexity underlying spectrum allocation decisions make the case for dramatic reform more compelling than ever before. Privatizing the management of spectrum promises the best use of this valuable resource for the American public.

American Opportunity

Although the Innovation Age is a worldwide phenomenon, its impact will not be uniform. Those places—cities, states, countries—most conducive to innovation will see more of it. In this regard, the United States should be in the forefront. To some Americans, this may seem more a curse than a blessing. They may lament the "destruction" more than they celebrate the "creativity." This view, though, ignores the fact that all members of society can benefit from economic progress that permits society to operate at higher plateaus of knowledge and understanding. It is also myopic, because it fails to note that America is well-positioned to spurt ahead of its nearest economic rivals. Again, the benefits to all Americans from widely available innovation will proceed directly from the availability of the innovative new services and products themselves, and indirectly from greater prosperity brought about by the better, cheaper goods and services produced by the new technologies.

Winner-Take-All Society

A conflicting view of the benefits of economic growth has recently been proposed by some economists concerned that the American economy is now rewarding high achievers inequitably compared to past years. In the *Winner-Take-All-Society*, Robert H. Frank and Phillip J. Cook theorize that this phenomenon is due to the increased importance of what they call "winner-take-all" markets.[66] Unlike typical economic markets, rewards in the winner-take-all variety are determined by relative, not absolute, performance. Also, "rewards tend to be concentrated in the hands of a few top performers."[67] The examples offered include superstar movie actors and top-level classical musicians, who earn compensation substantially above the very good actors and musicians one tier down. According to this theory, the telecom revolution, globalization of markets, falling transportation costs, and other factors cited here as driving the

Innovation Age, are creating many more winner-take-all markets.[68] Although Frank and Cook concede that winner-take-all markets produce substantial benefits (everyone can see the "best" actors and listen to the "best" musicians), they are concerned that these markets cause greater income disparity between the rich and everyone else, and are an inefficient use of society's resources. The latter, they say, is true for two reasons. First, too many "contestants" enter the winner-take-all markets because entrants overestimate their chances of winning and do not see the full costs to society of their choices. Second, these contestants undertake wasteful investments to increase their relative chances of winning. When undertaken by everyone, these investments do not change outcomes and do not benefit consumers.[69] For Frank and Cook, these inefficiencies justify the development of cultural and privately developed limits to check the excesses of winner-take-all markets. Frank and Cook also argue that the growth in importance of winner-take-all markets justifies increasing the progressivity of income tax rates, reorganizing our tax system to tax luxury consumption, reforming tort law, and undertaking a variety of other restructuring measures.[70]

Interestingly, the winner-take-all view recognizes that economic systems have undergone and are undergoing an important change. It fails, however, to appreciate the importance of the complexity and innovation that are arising out of the sources of this change. It is likely that Hayek would question many of Frank and Cook's policy recommendations. Although he would agree that many cultural traditions may have, at least partially, developed to adjust private incentives to meet larger societal goals, he would recommend caution to government officials attempting to restructure marketplace incentives.[71] In the Innovation Age, it is going to become even more difficult to predict where and when markets will have too many contestants and will exhibit wasteful investments. As Frank and Cook recognize, increasing the number of contestants can increase performance. They contend, however, that the foregone benefits of the contestants' alternative employment will be greater than any such improvement. Given the multitude of factors affecting decisions and the enormous complexity of their ramifications, Hayek would question whether anyone can really know this.

The "winner-take-all" perspective also fails to consider adequately the role of public goods. As explained earlier, public goods are those goods which, once produced, can be used by an unlimited number of consumers at little or no additional cost. Better communications, distribution, and marketing have greatly increased the return on the production of certain public goods. Many mass-market goods and services such as movies, books, music, and broadcast sporting events are public goods. Increasing the market size for public goods lowers the cost per customer. It also generates higher profits, which in turn attract more production. More importantly, the benefits of public goods often greatly exceed those captured by their creators.

As discussed in Chapter I, ideas and innovations are themselves public goods. As Schumpeter and Hayek stressed, markets spontaneously and continuously discover how the "pie" that is a country's economy can be made bigger. In the Innovation Age, this will come less from hard labor employed in making more widgets than from determining the best way to make widgets (spontaneous order) or in inventing a new kind of widget altogether (creative destruction). Once this knowledge is gained, the larger societal benefit to businesses, employees, stockholders, and consumers will be far greater than that achieved by the "innovator." Frank and Cook recognize this public-good effect of innovation:

> Consider, for example, the case of John Bardeen and Walter Brattain, the Bell Labs physicists who invented the transistor. . . . It has led to literally hundreds of billions of dollars' worth of expanded world output. Bardeen and Brattain themselves were not the patent holders for the transistor. But even if they had been, they could have captured only the most minuscule fraction of their product's value. . . . Economist Edwin Mansfield and his colleagues have estimated that innovations can be imitated for an average only 65 percent of their original cost, and that 60 percent of patented innovations are imitated within four years.[72]

Frank and Cook further note that innovation may be an important exception to the winner-take-all view:

> Because innovation often generates significant external benefits, we might expect market forces to call forth too little innovation. But markets for innovation also tend to have winner-take-all payoff structures, and this, as we have seen, tends to generate excessive activity. The net effect thus depends on which of these opposing tendencies is larger. If, as may often be the case, the divergence between price and social value dominates, then research and innovation may constitute an important exception to our claim that winner-take-all markets attract too many resources.[73]

This exception to Frank and Cook's description of the role of winner-take-all markets is critical. As discussed earlier, innovation is now assuming paramount economic importance: although the benefits of innovation have long outweighed their societal costs, the potential gains from innovation are becoming far greater than ever before. To the extent that winner-take-all markets foster innovation, they serve a useful role. The rising tide of innovation—the ultimate public good—will lift all boats.

Digital Derby

The U.S. and its citizens may be better positioned than other countries to exploit the telecom revolution and the Innovation Age. This country has a comparative advantage in this realm for several reasons. American culture cultivates individuality and is open to innovation. Our heterogeneous society, with its stress on entrepreneurship, has fostered the creation of highly innovative new companies. Furthermore, the U.S. has a long history of being open to new technologies,[74] and has been a leader in deregulating its telecommunications industries. As a result, as an analysis by McKinsey & Co. found, the U.S. telecom industry is often twice as productive as its counterparts in major European companies.[75] Finally, the country's democratic tradition and system of limited government with checks and balances help to make it more difficult for economic interests to use government to stop economic change.

The track records of other countries suggest disadvantages in these areas. For example, such factors as high telephone rates, and language differences that act to fragment the market, have held back the European Community countries' penetration into the online services market. In 1993 the EC countries, with a larger total population than the U.S., had fewer online databases—1,616 compared with the 3,057 run by U.S. companies. Adding in other types of electronic-information services, the European industry's $3.9 billion in revenue was dwarfed by the U.S. industry's $9.7 billion.[76] There are only ten computers for every one hundred workers in Japanese offices, as opposed to forty-two per one hundred in American offices.[77] In addition, 66 percent of U.S. desktop computers are linked to networks, whereas in Japan only 17 percent are so networked.[78]

Industrial policy in European countries and Japan has also produced a number of costly errors. A decade ago the French state telephone company began Minitel as an electronic substitute for printed phone books and gave away squat little desktop terminals to millions of French homes and businesses, charging only by the local call into the system. Now there are 6.4 million Minitel terminals around France, with users making approximately 85 million calls per month to check phone numbers, make theater and airline reservations, read news reports, and peruse the personals. But the service, based on television-related and not computer-based technology familiar elsewhere, has never taken root outside France because of the cost of the terminals. Similarly, Japan made a costly mistake in developing an analog HDTV system, which is now being superseded by digital technology developed largely through the cooperative efforts of private firms in the U.S. In 1994 a Japanese analog HDTV set cost $6,100. No wonder sales of these sets have stagnated.[79] Japan also erred in relying on industrial planning in developing its telecom network, failing to recognize the potential of fiber optics systems. As a result, by mid-1994 Japan's Nippon Telegraph & Telephone Corp. had managed to deploy only 30 percent of the total fiber optic cable laid by U.S. local phone companies. An obviously concerned senior official at Japan's Ministry of International Trade and Industry said, "We are now a second-class country and we may be downgraded to a third-class country if we don't change our policies."[80]

Some countries—such as the United Kingdom and New Zealand—have been active in promoting telecom competition and deregulation; in some areas they even lead the U.S. For example, many Britons are already able to obtain residential telephone service from their newly arrived cable company at prices 10 to 15 percent below that available from British Telecom. Often the cable companies provide better service as well.[81] New Zealand, too, has been in the forefront of opening its telecom market to competition and using market-based reforms to manage the electromagnetic spectrum. In both of these countries, however, the business impetus is often coming from American companies.

The benefit of America's competitive advantages is already evident. The microprocessor and software markets are dominated by American firms such as Intel and Microsoft. U.S. export growth has averaged 9 percent per year from 1987 to 1994, due in large measure to exports of competitive high-tech goods. This gain is three times that achieved by Japan and Germany.[82] Likewise, American consumers lead the world in their acceptance of the digital revolution. The U.S. has outpaced Japan in the percentage of households with computers—according to a 1994 Japanese source, ownership in the U.S. stood at 15.8 percent of households; Japan was at 5.7 percent. The U.S. also led in personal computers linked to local area networks (52 percent vs. 8.6 percent), the number of CD-ROM titles (4,000 vs. 1,000), and cable TV subscribers (57.21 million households vs. 1.08 million households). Americans use cellular phones four times as much as do the Japanese, even though service began four years earlier in Japan.[83]

There are other, more general indicators of the U.S. advantage. These include increased levels of investment generally and the development of highly computerized factories. Patent applications are up 35 percent over the last six years.[84] Output per worker has been growing at 2 percent per year since 1991, twice the rate of growth in the 1970s and 1980s.[85] Between 1992 and 1995 U.S. companies sharply increased capital expenditures—47 percent over those three years.

And whereas capital spending increased across the board between 1992 and 1995, computer investment in constant dollars increased by a breathtaking 184 percent.[86] Frank Raines, vice chairman of Fannie Mae, sees no end to the deployment of affordable new technologies: "We are convinced to a moral certainty that IT [information technology] investment will go on forever."[87]

The impact of these computer and telecom investment dollars is being felt in the manufacturing sector as well. Many U.S. companies, to the amazement of their global competitors, are using software to customize products produced at mass production speeds.[88] The roboticized systems that were deployed in Japan are now being outperformed by these newer and more versatile systems that better integrate automation with workers. It turns out that it is more efficient to use hard labor with software. Motorola, for example, uses this "soft manufacturing" approach to make highly customized pagers. In response to a salesman's detailed request, robots select the correct components and humans assemble the final product.

Many economists now believe that annual productivity growth in the U.S. has improved by a half point or more as a result of the overall investment boom that is now occurring. To put this improvement in perspective, if it had been sustained during each year of the last decade, total personal income in the U.S. would have increased by nearly $1.5 trillion.[89]

Notwithstanding these advantages, much more could be done in the U.S. to promote the telecom revolution. The telecom legislation of 1996 sets the stage, but as noted here, the three main areas of telecom regulation still require substantial reform. Regarding price regulation, the competitive pressures created by innovation will require wholesale changes in regulatory approach. Where regulation may be necessary due to a continuing monopoly, it should be flexible and designed to exploit the potential for competition and innovation. In the area of technical standards, the advantages of technical uniformity and innovation will be optimized in a competitive environment. Any federal subsidies designed to keep residential telephone rates low should be targeted to low-income consumers.

The management of the electromagnetic spectrum should be left principally to market forces. Who can seriously doubt that if Apple,

Intel, Compaq, and Microsoft had needed FCC spectrum to get to the market, the personal computer revolution would have been postponed? Few things will promote wireless computer applications and local telephone competition as much as driving down the cost of spectrum by several orders of magnitude. Fundamental spectrum reform would build on the U.S.'s current but perishable lead in the global digital derby.

A regulatory framework that slows innovation in this area, despite any plausible short run efficiency gains it might achieve, will foreclose a tremendous opportunity for the American people. The age-old lesson illustrated by the Trafalgar example at the beginning of this book—that "missing the boat" can have enormous repercussions—can now be multiplied by countless innovations just over the horizon. Rather than shackle digital communications to a regulatory paradigm designed for the analog era, our nation must now fashion telecom policies fitted for the Innovation Age.

Notes

Chapter One

[1]David and Stephen Horwath, *Lord Nelson, The Immortal Memory* (Viking, 1988), 262.

[2]David Horwath, *Trafalgar: The Nelson Touch* (Atheneum, 1969), 48, 43-49.

[3]Ibid., 48.

[4]"21st Century Capitalism," *Business Week,* 24 January 1994, 19.

[5]Friedrich Hayek, *Studies in Philosophy, Politics, and Economics* (University of Chicago Press, 1967), 162.

[6]Ibid, 261.

[7]Eamonn Butler, "Hayek's Prophecy—The Road to Capitalism," *Wall Street Journal*, 24 March 1992, A16.

[8]Hayek, 39.

[9]Joseph Schumpeter, *Capitalism, Socialism, and Democracy* (Harper and Brothers, 1942), 83.

[10]Ibid., 83.

[11]Ibid., 73-74.

[12]Ibid., 83.

[13]Ibid., 81-87.

[14]Ibid., 83.

[15]Ibid., 85.

[16]Ibid., 91.

[17]Edwin Mansfield, *Micro-economics, Theory and Application* (Norton and Co., 1970), 461-463; and F.M. Scherer and David Ross, *Industrial Market Structure and Economic Performance* (Houghton Mifflin Co., 1990), 660.

[18]Thomas Sowell, *Knowledge and Decisions* (Basic Books, Inc., 1980), 74.

[19]Ibid.

[20]"21st Century Capitalism," 17-19.

[21]Ibid., 40, 128.

[22]"The Keys to the Future," in "The Information Revolution," *Business Week,* Special Issue 1994, 60-63.

[23]For example, one leading economic text agrees with many of Schumpeter's points: The social cost of monopoly is modest. See Scherer and Ross, 678. Subsequent research has shown that productivity improvements are caused primarily by technical change and more skilled work force as opposed to increases in capital, etc. (ibid., 614). While there is no theoretical answer to the question of whether monopoly favors innovation, there is empirical evidence that innovation is correlated with R&D and that R&D is correlated with concentration (ibid., 642). Overall, a situation of mixed monopoly and competition is most likely to advance innovation (ibid., 660).

[24]For a list of industries where outsider firms produced major innovation, see Scherer and Ross, 653.

[25]Robert Hof, "Fortress Intel," *Business Week*, 22 August 1994, 28. See also, Peter Pitsch and David Murray, "A New Vision For Digital Telecommunications," *Hudson Briefing Paper No. 171*, December 1994, 4-5, 11, and throughout. Regarding Intel's competitive strategy, see Jim Carlton, "Intel Corp. Cuts Prices Up to 39 Percent For Pentium," *Wall Street Journal*, 3 February 1995, B10.

[26]Michael Rothschild, *Bionomics, Economy as Ecosystem* (Henry Holt, 1990).

[27]Peter Robinson, "Paul Romer," *Forbes ASAP*, 13 February 1996, 67, 70.

[28]"Economists for an Expanding Universe," *Business Week*, 16 May 1994, 64.

[29]"21st Century Capitalism," 19.

[30]Robert W. Fogel and Stanley Engerman, eds., *The Reinterpretation of American Economic History* (1971), 187, 194-95, 201.

[31]Scherer and Ross, 8-10.

[32]Schumpeter, 83.

[33]Scherer and Ross concluded in 1990 that "the deadweight welfare loss attributable to monopolistic resource misallocation in the U.S. lies somewhere between 0.5 and 2.0 percent of gross national product." Scherer and Ross, 667.

[34]For illustrative calculations, see Scherer and Ross, 31, 613; and Mansfield, 461-62.

Chapter Two

[1]Charles Babcock, "The Incredible Shrinking Company," *Computerworld*, 25 October 1993, 6.

[2]George Gilder, *Microcosm* (Touchstone, 1989), 46-49, 348-49.

[3]Even the local exchange industry, before the wide spread development of competition, is generally believed to have been improving by at least 5 percent per year. In setting the productivity rate for regulatory purposes, the FCC estimated that local exchange companies could achieve an annual productivity improvement of 5.3 percent. John R. Norsworthy, a nationally recognized expert on productivity measurement and consultant for AT&T, estimates that the productivity for the local exchange companies' interstate access services has grown at a rate of 7.3 percent. See Appendix and Comments of AT&T, In the Matter of Price Cap Performance Review for Local Exchange Carriers, CC Docket No. 94-1, 11 January 1996.

[4]Cynthia A. Beltz, "How to Lose the Race: Industrial Policy and the Lessons of HDTV," *American Enterprise* (May-June 1991): 26.

[5]Lucas Powe, Jr, *American Broadcasting and the First Amendment* (University of California Press, 1987).

[6]Hush-a-Phone, 20 F.C.C. 391, rev'd, 238 F.2d 266 (D.C. Cir. 1956), on remand, 22 F.C.C. 112 (1957).

[7]*The Telecom Revolution, An American Opportunity* (The Progress and Freedom Foundation Working Group, May 1995), 18.

[8]MCI, 60 F.C.C. 2d 25, app. B (1976), 62-64; and Petition of AT&T for a Declaratory Ruling and Expedited Relief, 67 F.C.C. 2d 1455 (1978).

[9]MCI v. FCC, 561 F.2d 365 (D.C. Cir. 1977), cert. denied, 434 U.S. 1040 (1978) (Execunet I); and MCI v. FCC, 580 F.2d 590 (D.C. Cir. 1978), cert. denied, 439 U.S. 980 (Execunet II).

[10] FCC Report, "Long Distance Market Share, Second Quarter 1995" (October 1995).

[11]See e.g., Amendment of Subpart L, Part 91, to Adopt Rules and Regulations to Govern the Grant of Authorizations in the Business Radio Service for Microwave Stations to Relay Television Signals to Community Antenna Systems, Second Report and Order, 2 F.F.C. 2d 725, 734.

[12]*The Telecom Revolution*, 22.

[13]J. R. Meyer, M. J. Peck, J. Stenason, and C. Zwick, *The Economics of Competition in the Transportation Industries* (Harvard University Press, 1969), 9.

[14]Ibid.

[15]Petition of MFS Communications Company, Inc. For a Notice of Inquiry and En Banc Hearing, Federal Communications Commission, filed 1 November 1993, 13.

[16]Average annual expenditures increased 5.6 percent to $658 from $623 in 1992. Since 1983, spending for service has doubled from $325, but percentage of total spending has remained about 2 percent for almost fifteen

years, according to the FCC's "Trends in Telephone Service" prepared by Industry Analysis Div. The report states: "The average American household now spends more on long distance service than on basic local service." *Communications Daily,* 13 February 1995, 5.

[17]Long distance companies say 45 cents of every dollar they charge goes to the Bell Companies in access fees, but that the Bells' real cost is only 9 cents of each dollar. Therefore, 36 cents of every dollar they collect, or over $14 of the average monthly bill of $39, goes to some subsidy. According to the Bell Companies, 70 cents of every dollar they collect in access fees is used to subsidize local rates. Using their estimates, over $12 of the average monthly long distance bill goes to a subsidy. See "Avoiding the Toll," *Wall Street Journal*, 20 March 1995, R6, R9.

[18]Ibid.

[19]Calvin Monson and Jeffrey Rohlfs, "The $20 Billion Impact of Local Competition in Telecommunications," mimeo, Strategic Policy Research, 1-3.

[20]Hatfield Associates, Inc., "The Cost of Basic Universal Service," July 1994, 4.

[21]Daniel Burstein and David Kline, *Road Warriors, Dreams and Nightmares Along the Information Highway* (Dutton, 1995), 336.

[22]John Keller, "AT&T's Cuts Are Just the First Shot in Telecom Wars," *Wall Street Journal*, 4 January 1996, B2.

[23]Hatfield, 4.

[24]Debra Wayne, "FCC Fields Panelist Comments on Universal Services Options," *RCR*, 22 April 1996, 8.

[25]Kenneth D. Boyer, "The Costs of Price Regulation: Lessons from Railroad Deregulation," *Rand Journal of Economics* 18 (1987): 408-416.

[26]P. Sheehan, "What Went Right," *Atlantic Monthly* (August 1993): 85.

[27]Ibid.

[28]H. Banks, "A Sixties Industry in a Nineties Economy," *Forbes*, 9 May 1994, 107.

Chapter Three

[1]Babcock, 6.

[2]Alan Deutschman, "The Managing Wisdom of High-Tech Superstars," *Fortune*, 17 October 1994, 197.

[3]William Tucker, "Emerging from the Productivity Paradox," *Washington Times,* 2 October 1994, A13.

[4]Ed Scannell, "The Secret History of the IBM PC `Gamble'," *Infoworld*, 12 August 1991, 47, 50.

[5]Ira Sager, "IBM's Sad Memories," *Business Week*, 21 February 1994, 39.

[6]Stratford Sherman, "The New Computer Revolution," *Fortune*, 14 June 1993, 56, 58.

[7]Ibid., 62.

[8]William M. Bulkeley, "Computer Firms Are Benefiting From Cost Cuts," *Wall Street Journal*, 6 July 1994, A4.

[9]"The Datamation 100," *Datamation*, 15 June 1993, 27.

[10]William M. Bulkeley, "Pact to Buy Cray Marks 'End of an Era' of Independent Supercomputing Firms," *Wall Street Journal*, 27 February 1996, B9.

[11]Market capitalization unadjusted for debt, using closing prices of 7 March 1996.

[12]Open architecture does not imply that the system is interoperable or networkable with other systems. Openness refers only to the ownership of the intellectual property that is embodied in the design. Either an open or a proprietary system may interoperate or network with another open or proprietary system.

[13]Linda Runyan, "A Retrospective: 40 Years on the Frontier," *Datamation*, 15 March 1991, 34-65.

[14]This is not intended to suggest that DOS and UNIX have similar internal operations; they are clearly different (batch processing vs. multiprocessing, file allocation table vs. partitioning, etc.). The outward appearance of the two, however, is too similar to be coincidental (both use directories and trees, many of the commands are the same, etc.).

[15]Microsoft Windows has become increasingly popular in recent years. Windows 3.1 is not an operating system; it must run "on top of " DOS. In contrast, Windows 95 is an operating system and the next step in the evolution of the operating system for computers using the Intel architecture.

[16]Nathan Brookwood, Principal Analyst for Microprocessors, letter to Peter Pitsch, *Dataquest*, 29 April 1996; and "Intel Market Share Rises," *New York Times*, 21 February 1994, C4.

[17]Elizabeth Corcoran, "IBM and Apple Near Macintosh Licensing Deal," *Washington Post,* 10 April 1996, F1; Jim Carlton, "Apple Will Be More Aggressive About Licensing Power Mac Line," *Wall Street Journal,* 19 September 1994, B6; "Spindler's Apple," *Business Week*, 3 October 1994, 88.

[18]Jim Carlton and Joan E. Rigdon, "A Bitter Apple Confirms It Will Go It Alone," *Wall Street Journal*, 9 February 1996, B1.

[19]Deutschman, 202.

[20]Richard N. Langlois, "External Economies and Economic Progress: The Case of the Microcomputer Industry," *Business History Review* 66 (Spring 1992): 29.

[21]This *mo*dulate and *dem*odulate function gives the device its name.

[22]Charles R. Morris and Charles H. Ferguson, "How Architecture Wins Technology Wars," *Harvard Business Review* (March-April 1993): 86-96.

[23]Sherman, 68.

[24]David Kirkpatrick, "Why Compaq Is Mad at Intel," *Fortune*, 31 October 1994, 172.

[25]"Intel Market Share Rises," D4.

[26]Kirkpatrick, 175.

[27]Robert Hof, "Fortress Intel," *Business Week,* 22 August 1994, 28; and Jim Carlton, "Intel Corp. Cuts Prices Up to 39 Percent for Pentium," B10.

[28]Sherman, 68-70.

[29]John Markoff, "Intel Forms Chip Pact With Hewlett-Packard," *New York Times*, 9 June 1994, C6.

[30]Paul B. Carroll, "Advanced Micro Devices Begins to Ship Chip Compatible With Intel's Pentium," *Wall Street Journal*, 27 March 1996, B6.

[31]Don Clark, "Intel's Near-Monopoly Faces Growing Attack From Perennial Gadfly," *Wall Street Journal,* 21 October 1994, A1.

[32]Carroll, B6.

[33]Nathan Brookwood, "Can Any Microprocessor Vendor Compete with Intel in 1996?" *Dataquest Predicts*, 12 February 1996, 5.

[34]Paul B. Carroll, "Cyrix Hopes to Chip Out Niche in Touch PC Business," *Wall Street Journal*, 6 March 1996, B4.

[35]John J. Keller, "AT&T Unveils Video Calling Between PCS," *Wall Street Journal*, 16 June 1994, A1, A5.

[36]"The U.S. Government vs. Microsoft," *The Technopolitics Report*, 1994, 1-2. This position of leadership led the Justice Department to file suit against Microsoft for "anticompetitive behavior." Microsoft offered steep discounts to hardware makers that shipped large quantities of computers. They also charged hardware makers a "per processor" fee for every PC they sold, regardless of whether the PC shipped with DOS. As a result of the settlement, Microsoft will desist from the processor fee immediately and will stop issuing contracts that require hardware makers to agree to buy large numbers of DOS packages in future years, regardless of whether the hardware maker needs the packages.

[37]Laurie Hays, "IBM in Talks to Cover Costs of New Linkup With Apple," *Wall Street Journal*, 20 October 1994, A3, A16.

[38]Charles McCoy, "Apple, IBM Kill Kaleida Labs Venture," *Wall Street Journal*, 20 November 1995, B8; and Bart Ziegler, "IBM, Apple, H-P to Disband Taligent, Big Layoffs Loom at Software Venture," *Wall Street Journal*, 1 December 1995, B8.

[39]Bart Ziegler, "Internet Software Poses Big Threat to Notes, IBM's Stake in Lotus," *Wall Street Journal*, 7 November 1995, A1.

[40]Audrey Choi and Don Clark, "Microsoft, MCI, Digital Plan a Venture to Develop Corporate Data Networks," *Wall Street Journal*, 8 April 1996, B6.

[41]Elizabeth Corcoran, "Java Jumps into the 'Net," *Washington Post*, 10 December 1995, H1.

[42]Don Clark, "Microsoft to Adopt Rival Technologies," *Wall Street Journal*, 8 December 1995, B2.

[43]Dinesh S. Dave and Kathy E. Fitzpatrick, "Price/Performance of Desktop Computers in the U.S. Computer Industry," *Information and Management* 20 (1991): 161-82.

Chapter Four

[1]Kenneth Labich, "The Geography of an Emerging America," *Fortune*, 27 June 1994, 88.

[2]Newt Gingrich, "Foreword," Alvin and Heidi Toffler, *Creating a New Civilization: The Politics of the Third Wave* (The Progress and Freedom Foundation, 1994), ix.

[3]Jim McConville, "Big Numbers Are Music to DBS Ears," *Broadcasting & Cable*, 25 September 1995, 46.

[4]"Interactive Ads May Be on Rise," *Broadcasting & Cable*, 17 October 1994, 52.

[5]Jared Sandberg, "Online Services' User Counts Aren't What They Seem," *Wall Street Journal*, 6 October 1995, B1.

[6]Kara Swisher, "Internet's Reach In Society Grows, Survey Finds," *Washington Post*, 31 October 1995, 1.

[7]Jared Sandberg, "On Line Population Reaches 24 Million in North America," *Wall Street Journal*, 30 October 1995, B3.

[8]Jared Sandberg and Bart Ziegler, "Internet's Popularity Threatens to Swamp the On-Line Services," *Wall Street Journal*, 18 January 1996, A1.

[9]See e.g., Mark Berniker, "Broadcast Networks Jumping Aboard the Internet," *Broadcasting & Cable*, 13 February 1995, 29.

[10]Charles R. Lee, "Liberty, Equality, Fraternity," speech, 14 July 1994, Allen and Co., Sun Valley Conference.

[11]Harry Jessell, "The Way Uphill: How Cable's Making It on the Infohighway; John Malone Interview," *Broadcasting & Cable*, 28 November 1994, 34, 44.

[12]"Cable Ready," *Wall Street Journal*, 20 March 1995, R8. According to John Malone, "Overbuilds are expensive; the upstream will be modest." See Anthony Ramirez, "A War Within a Single Wire," *New York Times*, 27 October 1993, C1.

[13]Mark Robichaux, "Despite Many Claims for 500-Channel TV, Long Road Lies Ahead," *Wall Street Journal*, 29 November 1993, A1. John Sie, chairman of pay-cable service Encore Media Corp., says that even if movies-on-demand take half the "recent hits" video rental market, it would generate only about $1 billion a year in cash flow for the entire cable industry.

[14]William Bulkeley and John R. Wilke, "Can the Exalted Vision Become Reality," *Wall Street Journal*, 14 October 1993, B1, B6d.

[15]Leslie Cauley, "PacTel Puts Off Interactive-Video Plans, Concentrating Instead on Wireless Cable," *Wall Street Journal*, 28 September 1995, A3.

[16]"Cable Ready," R8.

[17]"Don't Believe Hype About Future Markets," *Broadcasting & Cable*, 10 October 1994, 22. See also, "Viacom Consolidates Online, Interactive Activities," *Broadcasting & Cable,* 27 February 1995, 51.

[18]"Technologists Blame Technology for Interactive TV Delays," *Communications Daily*, 21 February 1995, 4-5.

[19]Ibid. See statement of Richard Peske, Interactive Digital Solutions' product marketing manager.

[20]"Disney TV Chief Sees Computers and TV Driving Convergence, Cites Obstacles," *Communications Daily*, 22 March 1995, 5.

[21]Ibid., 7.

[22]"Content Said to Trail Technology Development As Equipment Standards Vary," *Communications Daily*, 22 March 1995, 7.

[23]"Technologists Blame Technology," 4-5.

[24]Bart Ziegler, "Surveyed Consumers Want Interactive, But Won't Boost Cable-TV Bills for It," *Wall Street Journal*, 29 November 1994, B6.

[25]According to NBC Research Director Horst Stipp, if viewers expect interactive service to be on their computers, it should stay there, and the same for TV. Jeffrey Hamilton, General Instrument Director of Technology and Business Development, believes that interactive screens will evolve into modular terminals. See "Technologists Blame Technology," 4-5.

[26]Harry A. Jessell, "High Speed Modems A Top Priority," *Broadcasting & Cable*, 4 December 1995, 82.

[27]Rich Brown and Richard Tedesco, "Promise vs. Performance," *Broadcasting & Cable*, 29 April 1996, 10.

[28]"Technologists Blame Technology," 4-5.

[29]Bulkeley and White, B1.

[30]Ziegler, B6. Hewlett-Packard's home products division sponsored a random survey which found that the mean amount respondents would pay for interactive services and their basic cable-TV services was $29 a month—about the same level they currently pay for cable alone. Of the 28 percent of respondents who don't currently subscribe to cable, 47 percent said they would do so just to get interactive services. And of those who do get cable today, 67 percent said they would subscribe to interactive services.

[31]Leslie Cauley, "Phone Giants Discover the Interactive Path Is Full of Obstacles," *Wall Street Journal*, 24 July 1995, A1.

[32]Iris Cohen Selinger, "Cerritos Test Shows There's More to Learn About Interactive Television," *Advertising Age*, 25 October 1993, 25; and David Toberkin, "Analysts Skeptical of Cerritos Buy-rates," *Broadcasting & Cable*, 29 November 1993, 123.

[33]Bulkeley and Wilke, B6.

[34]"Technologists Blame Technology," 4-5.

[35]"An Electronic Roundtable," *Wall Street Journal*, 15 November 1993, R24.

[36]"What's Driving the New PC Shakeout," *Fortune*, 19 September 1994, 112.

[37]Mark Robichaux, "Despite Many Claims for 500-Channel TV, Long Road Lies Ahead," *Wall Street Journal*, 29 November 1993, A1.

[38]"Viacom to Sell Cable Systems, Target Software," *Communications Daily*, 20 October 1994, 3.

[39]Schumpeter, 83.

[40]"The Interactive Bazaar Opens," *Economist*, 20 August 1994, 49. John Hollis of Andersen Consulting thinks the biggest potential for interactive shopping is for food and household staples. Others point to surveys on consumer attitudes toward shopping and fear of crime as evidence that many consumers will rely on interactive shopping more generally. The chairman of "Shoppers Advantage"—which offers sales information about 250,000 products from hundreds of manufacturers with discounts up to 50 percent—claims his company can make money at half the 8-10 percentage point margins of warehouse clubs, the lowest cost conventional retailers. Nonetheless, in 1993 Americans bought only $200 million of goods from online services.

[41]Ibid.

[42]Rick Tetzeli, "Is Going On-line Worth the Money?" *Fortune*, 13 June 1994, 104, 108.

[43]Timothy L. O'Brien, "On-Line Banking Has Bankers Fretting PCS May Replace Branches," *Wall Street Journal*, 25 October 1995, A1.

[44]Brent Schlender, "What Bill Gates Really Wants," *Fortune*, 16 January 1995, 34-36, 50.

[45]See Eugene Volokh, "Cheap Speech and What It Will Do," *Yale Law Journal* 104 (1995): 1805.

[46]George Gilder, "Digital Darkhorse Newspapers," *Forbes ASAP*, 25 October 1993, 139.

[47]Thomas G. Krattenmaker and Lucas A Powe, Jr., *Regulating Broadcast Programming* (MIT Press and AEI Press, 1994), 42.

[48]"Advertisers See Big Benefits From Interactivity," *Communications Daily*, 28 January 1994, 5.

[49]Patricia Sellers, "The Best Way to Reach Your Buyers," *Fortune*, Special Issue (Autumn-Winter) 1993, 14-15.

[50]Ibid.

[51]The concept of the personal agent can take many forms. For example, consumers can already get updated sports, financial or headline news on pagers. See Linda Kay Sakelaris, "Information Services Via Pagers Could Be 'Touch-Down' Industry," *RCR*, 20 November 1995, 24.

[52]"A Giant Stirs," *Wall Street Journal,* 20 March 1995, R20.

[53]G. Christian Hill, "Electronic 'Agents' Bring Virtual Shopping a Bit Closer to Reality," *Wall Street Journal*, 27 September 1994, A1.

[54]Ibid.

[55]A QVC executive plans on enticing consumers to complete a "user profile," which will enable customers to "train" their own smart agent. QVC would charge an annual subscription fee of $40-$50, with chances for credits, rebates, and coupons with purchases. Mark Berniker, "QVC Banks on Smart Shopping, Q Online to Premiere in Early '95, Pending Deal with Online Leader," *Broadcasting & Cable*, 19 September 1994, 31, 35.

[56]Lawrence M. Fisher, "The Industry Cognoscenti Take a Stab at the Vision Thing," *New York Times*, 11 December 1994, 8. See also, Don Clark, "Novell Unveils Ambitious Plan to Sell Software That Links Variety of Devices," *Wall Street Journal*, 7 February 1995, B5.

Chapter Five

[1]Stanley Besen and Thomas Krattenmaker, Co-directors of the FCC Network Inquiry Special Staff, *New Television Networks: Entry, Jurisdiction, Ownership and Regulation*, vol. II, October 1980, 499-501.

[2]James Aley, "Give It Away and Get Rich!" *Fortune*, 10 June 1996, 91, 94.

[3]Andrew Kupfer, "Set-Top Box Wars," *Fortune,* 22 August 1994, 111.

[4]Motion of AT&T Corp. To be Reclassified as a Non-Dominant Carrier, filed with the FCC in CC Docket No. 95-427, 3. Since 1984 AT&T's share of toll revenues fell from nearly 90 percent to about 56 percent as of the second quarter of 1995. During this same period MCI's share increased from 5 percent to 18 percent and Sprint's share increased from 3 percent to 9 percent. See FCC Report, "Long Distance Market Share, Second Quarter 1995" (October 1995).

[5]Memorandum of Bell Atlantic Corporation et al., In Support of Their Motion To Vacate the Decree, *United States of America v. Western Electric Co., Inc., and AT&T* (No. 82-0192).

[6]John J. Keller, "AT&T Eagerly Plots A Strategy to Gobble Local Phone Business," *Wall Street Journal*, 21 August 1995, A1.

[7]David P. Reed, "The Prospects for Competition in the Subscriber Loop: The Fiber-to-the-Neighborhood Approach," FCC Staff Paper (September 1993). See also, Michael Marcus and Thomas Spavins, "The Impact of Technical Change of the Structure of the Local Exchange and the Pricing of Exchange Access: An Interim Assessment," The Telecom Policy Research Conference (September 1993).

[8]For an explanation of this economic principle see Robert Bork, *The Antitrust Paradox* (Basic Books, Inc., 1978), 229-31.

[9]Leland L. Johnson, *Toward Competition in Cable Television* (The MIT Press and The American Enterprise Institute for Public Policy Research, 1994), 29-43.

[10]Mark Robichaux, "Cable Modems Are Tested and Found To Be Addictive," *Wall Street Journal*, 27 December 1995, 13.

[11]Craig D. Dingwall, "The Last Mile: A Race for Local Telecommunications Competition Policy," *Federal Communications Law Journal*, 48 (1995): 105, 119.

[12]Even where there is no actual competition for particular customers, the potential for effective entry will serve as an important check on abuse (e.g., potential entry by cable systems could force competitive performance by phone companies). For example, the threat of entry by other network providers should drive prices down even if it only threatens a small percentage of a monopoly's business, and the threats could come from a multitude of directions. This is a lesson of the computer industry.

[13]Thomas W. Hazlett and Matthew L. Spitzer, *Public Policy Towards Cable Television*, vol. 1 of *The Economics of Rate Control* (The MIT Press,

1996). Forthcoming. See also, Robert W. Crandall and Harold Furchgott-Roth, *Cable TV, Regulation or Competition?* (Brookings Institution, 1996).

[14]Paul Farhi, "Cable Firms May Get FCC Aid on Rates," *Washington Post*, 21 October 1994, B1, B3. See also, Marylu Carnevale, "FCC Plan Would Allow Higher Charges As Cable TV Systems Add Channels," *Wall Street Journal*, 24 October 1994, B12.

[15]Ibid.

[16]Rich Brown, "Fingerhut Cans Shopping Service, Cites Changes in Business Climate for Decision," *Broadcasting & Cable*, 28 November 1994, 18.

[17]Regulatory restriction of the railroad industry's ability to adjust prices in response to the rise of truck competition in the 1950's severely undermined that industry. Calvin Monson and Jeffrey Rohlfs, "The $20 Billion Impact of Local Competition in Telecommunications," mimeo, Strategic Policy Research, 9.

[18]"Telecom Vote Signals Competitive Free-for-All," *Wall Street Journal*, 2 February 1996, B1; and Gautam Naik, "Bell Companies Ready a Charge Into Long Distance," *Wall Street Journal*, 5 February 1996, B4.

[19]In 1984, AT&T's local telephone operations were divided up among seven regional companies as the result of an anti-trust suit brought by the U.S. Justice Department. They include US West, Bell Atlantic, NYNEX, Southwestern Bell Corp., Bell South, Ameritech, and Pacific Telesis. Their territories do not overlap and cover the vast majority of the U.S. population.

[20]Burstein and Kline, 380.

[21]Peter Pitsch and David C. Murray, "The Rise and Fall of the Cable TV/Telephone Mergers of 1993," The Progress and Freedom Foundation, June 1996.

[22]That these mergers were not likely to generate anticompetitive effects was a view widely held by industry observers. See Peter Passell, "A Satisfactory Marriage," *New York Times*, 10 November 1993, D1.

[23]Burstein and Kline, 367-68.

[24]Amy C. Page, "Microsoft: A Case Study in International Competitiveness, High Technology, and the Future of Antitrust Law," *Federal Communications Law Journal* 44 (1994): 99.

[25]Gautam Naik, "Mergers Ring In a Daunting New Era for the Three Independent Baby Bells," *Wall Street Journal*, 22 April 1996, B1.

[26]Morris and Ferguson, 86-96.

[27]David Kirkpatrick, "Riding the Real Trends in Technology," *Fortune*, 19 February 1996, 60.

[28]Telecommunications Act of 1996, 104th Cong., 2d Sess., H.R. Conf. Rept. 458, §629 and legislative history.

[29]Elizabeth Corcoran, "A 'Battle of the Box' Is Brewing," *Washington Post,* 15 October 1994, D1.

[30]Milton Mueller, "Universal Service In Telephone History, A Reconstruction," *Telecommunications Policy* (July 1993): 352.

[31]Telecommunications Act of 1996, §254(b)(3).

[32]Kenneth Gordon and John Haring, "The Effects of Higher Telephone Prices on Universal Service," FCC Staff Report, March 1984, 6.

[33]Alexander Belinfante, "Telephone Subscribership in the United States," Industry Analysis Division of Common Carrier Bureau, FCC (December 1995).

[34]Ted Bunker, "Can U.S. Afford Phone Subsidies? Universal Service Laws Skew Markets, Hurt Poor," *Investor's Business Daily*, 11 January 1995, A1.

[35]Gordon and Haring, 7.

[36]Economists John Panzar and Steven Wildman explain that "[g]iven large economies of scale and density, it should not be surprising that telephone costs are much higher in rural than in urban areas. The monthly average loop cost per access line is $45.50 for local exchange companies with less than 1,000 access lines, but only $18.90 for Tier 1 [the largest] companies, which have on average 1,284,500 access lines." Panzar and Wildman also illustrate how this externality might operate in a rural and urban setting. They posit a hypothetical situation in which 136,000 urban householders are better off contributing $48,000 per month to the cost of providing rural service to 4000 rural households, because the addition of the 4000 households increases the value of the network to each of the urban households by $.80 per month. (The product of 136,000 and $.80 equals $108,800, which greatly exceeds $48,000.) John Panzer and Steven Wildman, "Competition in the Local Exchange: Appropriate Policies to Maintain Universal Service in Rural Areas," mimeo, Northwestern University (1994), 7, 10, and 16.

[37]See Gordon and Haring. Also Crandall found that real price increases for telephone access and usage of roughly 30 percent during the 1980s reduced telephone subscription at the lowest levels (incomes under $10,000) by about 3 percentage points and that the effect was less for lower middle income households ($10,000 to $15,000). See R. W. Crandall, *After the Breakup, U.S. Telecommunications in a More Competitive Era* (The Brookings Institution, 1991), 111-12. Crandall's results are consistent with other research that estimated that a doubling of local telephone rates would

reduce overall subscription rates by 5 percent and rates for low income households by 11 percent. See R. E. Park and B. M. Mitchell, "Local Telephone Pricing and Universal Telephone Service," R-3724-NSF (Santa Monica, CA: Rand Corporation, June 1989).

[38]J.P. Fuhr, Jr, "Rural Telephony Since Divestiture," *Review of Industrial Organizations* 6 (1991): 89-95. Analyzing monthly residential rates for 483 rural telephone companies since the divestiture, Fuhr found only slight increases in the real residential rates.

[39]Hausman, Tardiff and Belinfante, "The Effects of the Breakup of AT&T on Telephone Penetration in the United States," *AEA Papers and Proceedings* (May 1993): 182.

[40]Crandall, 145.

[41]Fuhr, 89.

[42]Hausman, Tardiff, and Belinfante found that even low income (lifeline) customers pay a substantial portion of their monthly bill for toll services. See Hausman, 182.

[43]Ibid.

[44]See Telecommunications Act of 1996, §254.

[45]Daniel Pearl, "Despite Deregulation, Rural Phone Subsidies Are Likely to Survive," *Wall Street Journal*, 30 November 1995, A1.

[46]Hatfield, 4.

[47]See *The Telecom Revolution*, 63.

[48]P. Cain and J. MacDonald, "Telephone Pricing Structures: The Effects on Universal Service," *Journal of Regulatory Economics* 3 (1991): 293-308.

[49]Park and Mitchell.

[50]Peter Pitsch and David Teolis, "Not Mutually Exclusive," *Public Utilities Fortnightly*, 15 March 1996, 29.

[51]As stated by Hausman, Tardiff, and Belinfante, the subscriber line charge accounts for approximately one-third of the average price of measured-rate basic access. Accordingly, using the own-price elasticity, increases in the subscriber line charge would lead to a predicted decline in penetration. They determined that lower long-distance rates had a positive effect on penetration of approximately three times the magnitude of the increase in basic exchange access prices. See Hausman, 182-3.

[52]"A Legislative Agenda For Telecommunications," *Issues in Science and Technology* (Fall 1993): 59.

[53]Telecommunications Act of 1996, §254(c)(1).

[54]Ibid, §254(h).

[55]Burstein and Kline, 393-94. See also, *Communications Daily*, 6 June 1994, 6.

[56]David Kirkpatrick, "How PCs Will Take Over Your Home," *Fortune*, 21 February 1994, 101.

[57]Mark Berniker, "Internet Begins to Cut Into TV Viewing," *Broadcasting & Cable*, 6 November 1995, 113.

[58]Blayne Cutler, "The Fifth Medium," *American Demographics* (June 1990): 61.

[59]Ibid.

[60]See Bell Atlantic's apparent success in Union City, New Jersey, "Union City Interactive Multimedia Education Trial," Education Development Center, Inc.

[61]Suzanne Alexander, "The Digital School," *Wall Street Journal*, 15 November 1993, R17.

[62]Jeffrey Rohlfs, Charles Jackson, and Tracy Kelly, "Estimate of Loss to the United States Caused by the FCC's Delay in Licensing Cellular Telecommunications," mimeo, National Economic Research Associates (November 1991).

[63]George Calhoun, *Digital Cellular Radio* (Artech House, 1988).

[64]Evan Kwerel and John Williams, "Moving to a Market for Spectrum," *Regulation*, 7 February 1996, 59.

[65]"FCC Hits $15 Billion Mark in Total Net Auction Revenues," *FCC Daily Digest News Release*, 7 February 1996. This analysis needs to be updated with the results from other recently completed auctions. See "FCC's C Block Spectrum Auction Ends After Raising $10.2 Billion," *Communications Daily*, 7 May 1996, 1; "The FCC's Multipoint Distribution Service Auction Closes," *FCC Daily Digest News Release*, 29 March 1996; and "The FCC's 900 MHz Specialized Mobile Radio Auction Closes," *FCC Daily Digest News Release*, 15 April, 1996.

[66]Robert H. Frank and Philip J. Cook, *Winner-Take-All-Society* (Free Press, 1995).

[67]Ibid., 24.

[68]Ibid., 45-60.

[69]Ibid., 125.

[70]Ibid., 211-231.

[71]In describing the complex nature of the market's spontaneous order, Hayek writes the following: "It turns out to be an instance of a general method of indirectly creating an order in situations where the phenomena are far too complex to allow us the creation of an order by separately putting each element in its appropriate place. It is a sort of order over the particular manifestation of which we have little control, because the rules which determine it determine only its abstract character, while the detail

depends on the particular circumstances known only to its individual members. It is therefore an order which we cannot improve upon but only disturb by attempting to change by deliberate arrangement any one part of it." Hayek, 92.

[72]Frank and Cook, 120-21.

[73]Ibid.

[74]"Why Are We So Afraid of Growth," *Business Week*, 16 May 1994, 72.

[75]Ibid.

[76]Richard L. Hudson, "Incommunicado, Language Barriers and High Phone Charges Have Slowed the Spread of On-line Services in Europe," *Wall Street Journal*, 15 November 1993, R19.

[77]Burstein and Kline, 301.

[78]Ibid.

[79]"Do Not Adjust Your Set," *Japan Scope* (Summer 1994): 29.

[80]David P. Hamilton, "Big Fiber-Optic Project Is Private Sector's Job, Japan's Reformers Say," *Wall Street Journal*, 15 August 1995, A1.

[81]"Down the Line," *Economist*, 26 November 1994, 76.

[82]"Why Are We So Afraid of Growth," 72.

[83]Hamilton, A1.

[84]"Why Are We So Afraid of Growth," 65.

[85]Joseph Spiers, "The Most Important Economic Event of The Decade," *Fortune*, 3 April 1995, 40.

[86]Ibid., 33.

[87]Ibid., 34.

[88]Gene Bylinksy, "The Digital Factory," *Fortune*, 14 November 1994, 93.

[89]Spiers, 40.

Index

A

B

C

D

E

F

G

H

I

J

K

L

M

About the Author

Peter K. Pitsch is an adjunct fellow at Hudson Institute and Director of the Competitiveness Center's Telecommunications Project. He is also a consultant specializing in telecommunications law and economics and has been an adjunct law professor at Georgetown University Law Center. He has represented telephone, cable, cellular, and customer interests on a wide variety of communications issues including price caps, subscriber line charges, tariffs, broadcast ownership, financial interest and syndication, syndicated exclusivity, and comparative renewal. Prior to founding his own consulting firm, Pitsch Communications, in 1989, he served as Chief of Staff to Chairman Dennis R. Patrick at the Federal Communications Commission (1987-1989) and Chief of the FCC's Office of Plans and Policy (1981-1987). He received his J.D. in 1976 from Georgetown University Law Center and a B.A. in Economics from the University of Chicago in 1973. He also worked as a senior attorney at Montgomery Ward, Inc. and as an attorney-advisor to Commissioner Calvin J. Collier of the Federal Trade Commission. Mr. Pitsch has delivered numerous speeches to industry groups and has written articles on important communications issues for leading newspapers and national magazines.

About Hudson Institute

Hudson Institute is a private, not-for-profit research organization founded in 1961 by the late Herman Kahn. Hudson analyzes and makes recommendations about public policy for business and government executives, as well as for the public at large. The institute does not advocate an express ideology or political position. However, more than thirty years of work on the most important issues of the day has forged a viewpoint that embodies skepticism about the conventional wisdom, optimism about solving problems, a commitment to free institutions and individual responsibility, an appreciation of the crucial role of technology in achieving progress, and an abiding respect for the importance of values, culture, and religion in human affairs.

Since 1984, Hudson has been headquartered in Indianapolis, Indiana. It also maintains offices in Washington, D.C.; Madison, Wisconsin; and Brussels, Belgium.

About The Progress & Freedom Foundation

The Progress & Freedom Foundation is a private, non-partisan idea center established in 1993. Its goal is to ceate a positive vision of the future founded in the historic principles of the American Idea. It brings together a diverse group of thinkers and policy experts and shares their work with the American people through seminars, conferences, publications, and electronic media of all forms. Supported by tax deductible donations from corporations, foundations, and individuals, The Progress & Freedom Foundation does not engage in lobbying activities or take positions on legislation.